Springer Transactions in Civil and Environmental Engineering

More information about this series at http://www.springer.com/series/13593

Fumihiko Seta • Arindam Biswas
Ajay Khare • Joy Sen
Editors

Understanding Built Environment

Proceedings of the National Conference on Sustainable Built Environment 2015

Springer

Editors
Fumihiko Seta
Department of Urban Engineering
The University of Tokyo
Tokyo, Japan

Ajay Khare
School of Planning and Architecture
Bhopal, MP, India

Arindam Biswas
Department of Architecture & Planning
IIT Roorkee
Roorkee, Uttarakhand, India

Joy Sen
RCG School of Infrastructure Design and Management
Department of Architecture and Regional Planning
IIT Kharagpur
Kharagpur, WB, India

ISSN 2363-7633 ISSN 2363-7641 (electronic)
Springer Transactions in Civil and Environmental Engineering
ISBN 978-981-10-2136-7 ISBN 978-981-10-2138-1 (eBook)
DOI 10.1007/978-981-10-2138-1

Library of Congress Control Number: 2016955548

Printed on acid-free paper

This Springer imprint is published by Springer Nature
The registered company is Springer Nature Singapore Pte Ltd.
The registered company address is: 152 Beach Road, #22-06/08 Gateway East, Singapore 189721, Singapore

Preface

Built environment is regarded as a discipline which discusses human intervention that reshapes natural settings on earth. It is desired that human intervention is minimal and can coexist with nature, but the reality is somewhat different. At this moment when the global future is at stake with increasing population, random urbanization, threat of climate change, social and economic inequality, poverty, deprivation and violence, the response from human race is mediocre. The issues have been raised in several platforms but not in the magnitude which it deserves. Many times the issues are forgotten in the aftermath or even in the absence of a significant platform. But we were determined to follow up the deliberations of the august gathering for Sustainable Built Environment 2015 conference at the sprawling campus of IIT Roorkee during April 10–12, 2015. The aim was to garner more meaningful contribution towards the society and nature, which anchors our very existence. We have walked the extra mile during the difficult journey with lots of discussions, debate and argument and to finally come in a position to publish a document, which is a living manifestation of a divine bliss to enhance knowledge, capability and creativity among the readers. We hope that the ethos of this publication and every word printed in this book will further spread among the society to enhance more knowledge and built innovative ideas to achieve a sustainable and resilient built environment.

India is one of the countries which are at the centre stage of evolution and has all the plights of present global development concerns. Her population is 1.2 billion and rising fast; half of her population is going to be urban by 2030; 22 % of her population survives below $1.25 a day; it's very much prone to natural disaster and recognized as one of the countries to be severely affected due to climate change; and her citizen faces social, economic, gender, religious and political violence. Within this backdrop, the point of discussion for planners and architects needs to be responsive and socially responsible. And so is the sole aim of this book. It discusses Indian context in various dimensions of growth and development and our understanding towards built environment. Spatial patterns are influenced by various contextual parameters which in turn reflect in shaping the image of the city and its specific social, cultural and architectural expression. Long enduring division between urban and rural region is gradually fading away, at least with the emerging extent of communication, networking and globalization. But the economic disparity still exists in developing countries like India. However, from the past couple of decades or so there is increasing urge in India to connect habitable environment more to achieve better quality of life. This is the era which experienced much focused attention towards comprehensive planning and designing built environment than ever before. Even legislative framework to back up the development process improved with globalization and decentralization. Urban reform, efficiency in service delivery, financing development projects, infrastructure development and resource utilization with smart initiatives are few of the hallmark initiatives in the recent times.

The editors ensure that the book critically endures every word and see that all the efforts by the contributors become audible to a greater society. The purpose of this approach is to make it more holistic rather focusing on only one dimension of built environment. We have piece together all the evidences from empirical study to theoretical argumentation and encourage critical views from the contributors. Therefore, it not only records the development process but also showcases careful critique of the focus areas of individual contribution. This book covers architectural advances and sustainable urban regeneration, which have profound impact in understanding built environment. It presents the spatial context which hosts human to survive, enrich, contribute and disseminate. Architectural expressions and urban regeneration are interrelated and have broader implication in cities. But somehow it is also connected in the very root of built environment. This book wishes to touch upon the whole network from small to large that connects built environment and human behaviour. The book receives overwhelming response from the contributors across the country, and they can combine all their ideas, thoughts and contribution in two volumes. A total of 26 contributors have made their marks in this book. Divided into two parts, the book presents their works (14 in part 1 & 8 in part 2) in a precise manner.

Bunkyo-ku, Tokyo, Japan — Fumihiko Seta
Roorkee, Uttarakhand, India — Arindam Biswas
Bhopal, MP, India — Ajay Khare
Kharagpur, WB, India — Joy Sen

Acknowledgement

Acknowledgement is conveyed with a sincere thanks to all of our family members and friends. A special thanks to Prof. Pradipta Banerji (Director IIT Roorkee) and Prof. Ila Gupta (Head, Department of Architecture and Planning, IIT Roorkee) for their encouragement to accomplish the outcome. The students, teachers and staffs of IIT Roorkee deserve all the accolades for being the strength of this platform and the book. Their hard work paved the way for today's culmination. We would particularly like to thank Ms. Rohita Sharma, the then research scholar at the Department of Humanities and Social Science IIT Roorkee, for serving in the conference secretariat, and Ms. Ankita Choudhary, the then MURP student of IIT Roorkee, who carried out secretarial assistance to the editors. At this very occasion, the financial supporters who provide the realistic strength need a special mention from us for enabling this event. We are extremely grateful towards our editor and the rest of the staff at Springer for their patience, continuous guidance, help and suggestion.

We have tried whole heartedly to contact and acknowledge copyright owners. If something or someone is left out, the authors and publisher would be extremely pleased to have any errors or omissions brought to their attention. We will make all the necessary corrections and have them published at a later printing.

Contents

About the Editors

Dr. Fumihiko Seta is currently an associate professor of the Department of Urban Engineering (DUE), Faculty of Engineering at the University of Tokyo. He was born in 1972 in Tokyo, graduated in the University of Tokyo in 1995 and worked at the University of Tokyo from 1998 to 2005 and Osaka City University from 2005 to 2012. His major is urban planning and regional development.

Dr. Arindam Biswas is assistant professor at the Department of Architecture and Planning. He holds a doctorate in Urban Planning from the University of Tokyo. He is the recipient of the DAAD Fellowship and the MEXT Scholarship. His research specializations are on urban and regional planning, inclusive growth, urban policy and governance, resilience and futuristic city planning. He is presently serving as a reviewer for *Development in Practice*, *World Development* and the *Journal of Architectural Engineering* (ASCE).

Professor (Dr.) Ajay Khare founder-director of School of Planning and Architecture, Bhopal, is a Fulbright Fellow and Charles Wallace Fellow. He was the first Indian recipient of Berkeley Teaching Fellowship. Professor Khare served as a member of the Planning Commission working group on Technical education for 12th plan of India. He is a member of the National Executive Committee of the International Council on Monuments and Sites (ICOMOS, India).

Professor (Dr.) Joy Sen is head of RCG School of Infrastructure and Design, IIT Kharagpur. His areas of research are Community and Regional Planning Analyses and Programming and Architecture and Planning. He holds a Ph.D. from IIT Kharagpur and a Master's from Iowa State University. He is currently involved in SandHI project sponsored by MHRD. He is a reviewer of the *Journal of Architecture* and associate editor of *CSVTU Journal*. He has authored five books on Indian Heritage and Iconographic Documentation.

Part I
Land, Housing, and Real Estate

Chapter 1
Accessibility to Health Infrastructure in Hill Settlements: A Case of Shimla

Adwitiya Patro

Abstract Urban areas are spreading their extents rapidly across different geographic conditions and simultaneously affecting the hill topographies. This rampant development brings inequity in terms of access to services, and accessibility being a major factor affecting the livelihoods need special attention. Indian Cities generally has concentration of services in their core, bringing friction for residents living across the city, when accessibility is concerned. In hill cities, the accessibility parameters take account of multiple factors and majorly, topographic factors. Study enquires the relationship between residential choice and accessibility to health amenities for city of Shimla (H.P). As evident in hills, degraded accessibility in such areas can lead to multiple social disorder and sense of insecurity, along with added economic cost. Study measures the network accessibility analysis for health infrastructure in hill areas using FlowMap® and GIS-based tools. In this study it has been tried to analyze how the different residential areas of Shimla perform in terms of network access to hospitals. Which can certainly be used in multiple urban planning decision and in diverse sectors.

Keywords Accessibility • FlowMap® • GIS • Health infrastructure • Shimla

1.1 Introduction

Effects of unplanned urban growth is evident in all major hill cities across India. Indian hills contain around 10 % of India's total population, witnessing an unexpected growth largely due to industrialization which induces urbanization into these fragile ecosystems. Of all the hill settlements, once called Queen of Hills, Shimla has lost its pristine glory with rapid urbanization in recent times. As per the article

A. Patro (✉)
CEPT University, Ahmedabad, India
e-mail: adwitiya@live.com

F. Seta et al. (eds.), *Understanding Built Environment*, Springer Transactions in Civil and Environmental Engineering, DOI 10.1007/978-981-10-2138-1_1

by Manjeet Singh, a city earlier designed for 25,000 inhabitants now caters to seven times the number stated above and putting tremendous pressure to urban systems.

Individual travel choice due to lack of access, affect mode of travel which in turn lead to induced poverty as we see in the case of women users from economically backward classes in Delhi's case (Anvita and Geetam 2006).

Thus, as much as the physical infrastructure development in terms of technology and efficiency in urban transport, there is a high need of impetus to be put to the matter that "accessibility" is what we are aiming for, with all these investments.

1.2 Shimla: An Introduction

Shimla is the capital town of Himachal Pradesh with an average elevation of 2205 m (7234 ft) from mean sea level with an area of 25 km^2, having a population of approximately 1.7 lakhs, with a density of 120 per sq.km. Shimla lies in the lower ranges of the Himalayas in its southwest range with a coordinate of 31.61 ON and 77.10 OE (UNDP 2003). The city stretches itself in the east-west direction with a length of around 9.2 km. Shimla is embodied around seven hills, the highest being Jakhoo hills with an elevation of 2454 m. This geographical setup possesses hindrance for regional connectivity with the city. With rapid growth of population and lack of proper transport infrastructure in city and around, this issue of connectivity gets deteriorated progressively. Hospitals like Rippon Hospital and Walker Hospitals were constructed in 1885 and 1906, respectively, which are prominent in the city.

Being in hill topography, Shimla possesses a threat in terms of fragile ecosystem through the rampant development it is going under. So to ensure planned growth and development, GoHP[1] constituted Shimla Planning Area (SPA), as it is mentioned (Sekhar 2011) around 82 % of the population of SPA comes under SMC.[2] Shimla a city with tourism as its primary economic activity, also known for its role in administration, education, heritage city, etc.

As Sashi Sekhar (2011) highlights in his research, Shimla has experienced massive urban pressure as there has been ribbon growth along the roads. Also it has been seen that this population pressure is making built growth in steep slopes too which are unsustainable in many ways. As per the study (Sekhar 2011), he highlights that built-up area except roads in Shimla has grown by 7 %.

Unregulated development in turn affects the quality of life. With JNNURM, efforts have been made to improvise the public transport and housing infrastructure, seem to be futile with no drastic change happening until now (Carol 2012). In a recent study done by Mazta and Thakur (2012), it has been highlighted that due to the lack of accessibility to health infrastructure, villagers in Himachal Pradesh face a lot of health issues.

[1] Government of Himachal Pradesh.

[2] Shimla Municipal Corporation.

Looking at the scenario, study wants to focus on the network accessibility to health infrastructure in Shimla city.

1.3 Health Infrastructure in Hill Areas

As in the study done by Dev Nathan (2004), it has been highlighted that the level of health infrastructure available in hill is significantly poor in comparison to plains. Investments and standards are prepared but hardly has it seemed that factors like accessibility are taken care of. In the research done by Sapkota (2015), it has been found that with increasing time to health infrastructure, access to it decreases which eventually affects human health.

Since health infrastructure varies in scale, the study only considers public hospitals, large-scale private hospitals, and PHCs[3] in correspondence to standards prescribed by URDPFI[4] guideline (as highlighted in Table 1.1), which has been used for studying the requirement and capacity of each health facility.

1.4 Accessibility: Literature

Accessibility has a plethora of meaning and is derived from its Latin root, i.e., "accessus" which means "a coming near." But what we are talking in our context can be defined as "the ease with which services are approached within stipulated amount of time" (which can be a measure of distance). Further according to Geurs and van Eck (2001), accessibility is basically the effort or amount of time that is

Table 1.1 URDPFI standards for health infrastructure

Healthcare facilities			
S/No.	Category	Population served per unit	Distance between two facilities (KM)
1	Dispensary	2500	2–4
2	Health Sub-Centre	3000	2–4
3	Family Welfare Centre	5000	5–10
4	Maternity Home	15,000	5–10
5	Nursing Home	15,000	5–10
6	Primary Health Centre	20,000	16–20
7	Hospital (200–250 beds)	80,000	16–20

[3] Primary Healthcare Centre.

[4] Urban and Regional Development Plans Formulation & Implementation Guidelines.

given for a destination to be reached by a person or activities that can be reached from a particular location.

1.5 Methodology of Accessibility Analysis

For the study we have used activity-based accessibility measures as both space and time factors are taken care of in the study. In this the use of GIS tools and FlowMap®[5] has been used for modeling the areas. Before the analysis, first the data has been generated through many sources.

1.5.1 Road Network

To map the road network, 2.5 m Cartosat-1 imagery was used and elevation points has been given as an attribute. This consists of NH22, NH88, state highways, and other major roads. Local roads have not been captured due to the need of intensive surveying and data collection. The research intends to develop a methodological input, accordingly datasets have been created. As local roads could not be captured, an assist network has been developed with a 150 m grid. Then the values of elevation have been captured from DEM,[6] so as to calculate the slope for every road segment. Different network speeds have been assumed and then accordingly time has been computed with the following formula:

$$T = D * S^{-1}$$

where T = time
D = distance
S = speed

For the network speeds, assumptions have been taken into consideration (as Table 1.2), having a coherence with IRC[7] 73-1980, and assist network[8], primarily as a pedestrian network, was assumed and adopted into the scope of present analysis. Accordingly a city road network hierarchy map (as Fig. 1.1) has been generated in ArcGIS.

[5] Geographical information system and FlowMap® are registered software of the University of Utrecht which is an open ware and helpful for Land use-transport scenario analysis.

[6] Digital Elevation Model.

[7] Indian Road Congress.

[8] Assumed to be a pedestrian network primarily for which generic walking speed is considered, kindly refer to Table 1.2.

Table 1.2 Assumed network speeds

Road type	Speed assumed (kmph)
NH	40
SH	30
City	20
Assist	3

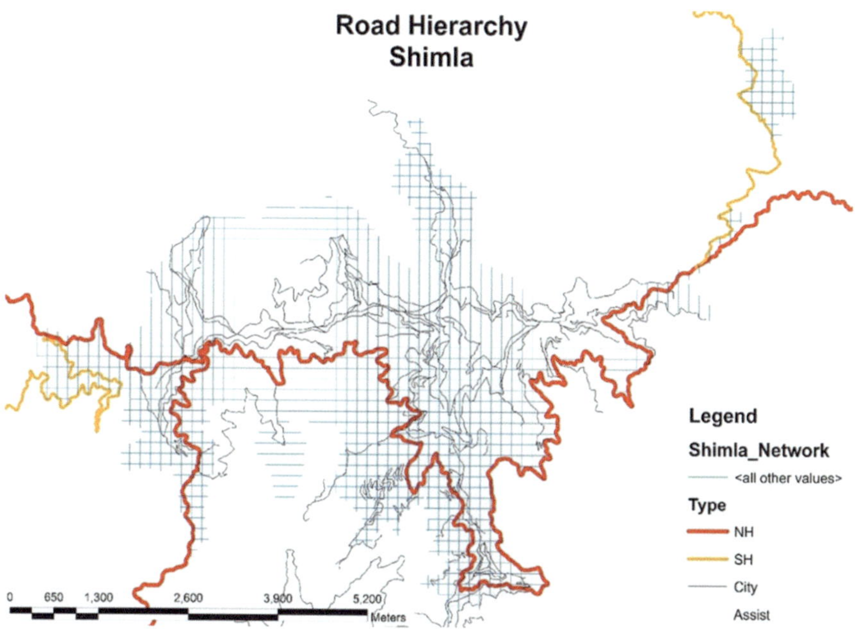

Fig. 1.1 Road hierarchy

For the required analysis, slope adjustment factor has been taken into consideration, so as to include factor of resistance into the analysis. For slope adjustment factor, formula has been adopted from a research by Mike Price (n.d.):

$$\text{Slope Adjustment Factor}(D) = 1 + (\text{Slope in Percentage}/10)^2$$

where

$$\text{SlopeinPercentage} = ((\text{EndElevation} - \text{Start elevation})/\text{Length of Segment}) * 100.$$

Then network resistance has been computed with the following formula, and in it slope adjustment formula acts as a decay function, thus increasing the resistance in exponential manner:

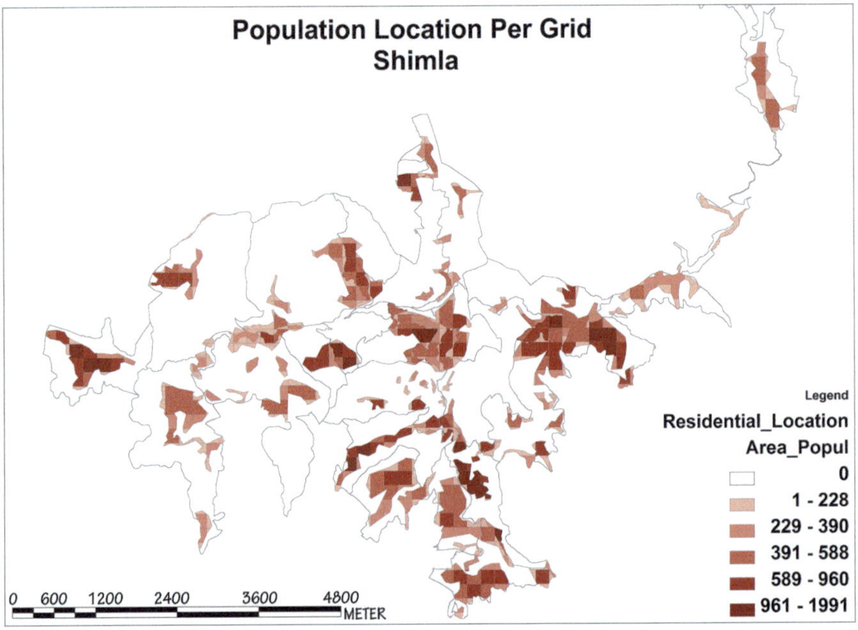

Fig. 1.2 Residential location and population distribution

$$\text{NetworkResistance} = T * D.$$

This network resistance is computed for every segment and becomes an input into network analysis.

1.5.2 Residential Cluster

For ease of analysis, only residential clusters have been identified with Cartosat-1 imagery, and then population has been given as input from Census 2011.[9] Also the grid of 200 m has been created within the demarcated residential clusters (as Fig. 1.2).

1.5.3 Hospital Locations

Mainly public hospitals have been identified of different scale, and their capacity as per URDPFI guidelines has been given as input into analysis, with their geolocation embedded in it and hereby highlighted (as Fig. 1.3).

[9] Census 2011 for Shimla urban area.

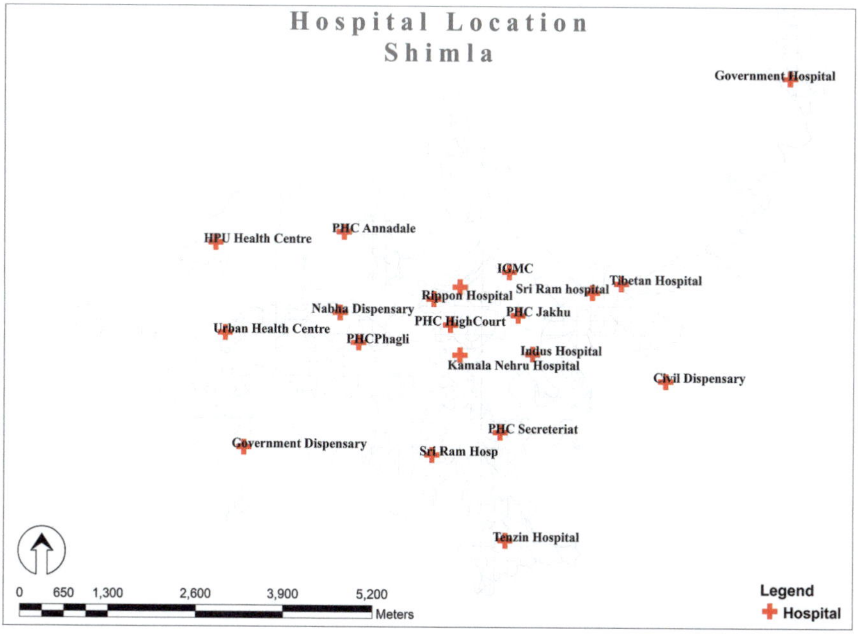

Fig. 1.3 Hospital locations

1.6 Analysis

1.6.1 Proximity Analysis

Proximity analysis for residential location to health infrastructure locations has been done with computed network speed[10] for time period of 10, 20, 30, and 45 min (as Figs. 1.4, 1.5, 1.6, and 1.7, respectively). The following are the output; lightest hue depicts least or no accessibility and dark blue shows the highly accessible zones[11] for which FlowMap® Software has been used.

1.6.2 Catchment Area Analysis

Catchment area analysis for each location of health infrastructure, with attributes of residential clusters and health infrastructure (as Figs. 1.8 and 1.9), according to URDPFI guidelines, has been studied. Thereafter gradation was done according to percentage of residents getting access within a fixed time period.

[10] Network speed has been adjusted for each of the segments which includes slope adjustment factor that has been highlighted beforehand.

[11] The yellow zones are those where there is no accessibility to health infrastructure, whereas the dark blue areas are those where one can at least have access to a city-level hospital.

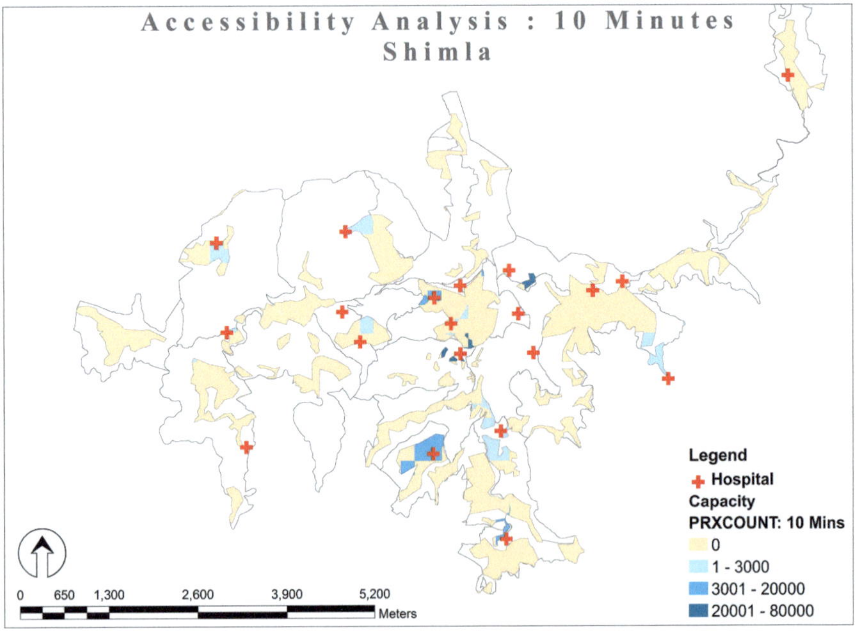

Fig. 1.4 Proximity analysis 10 min in network speed

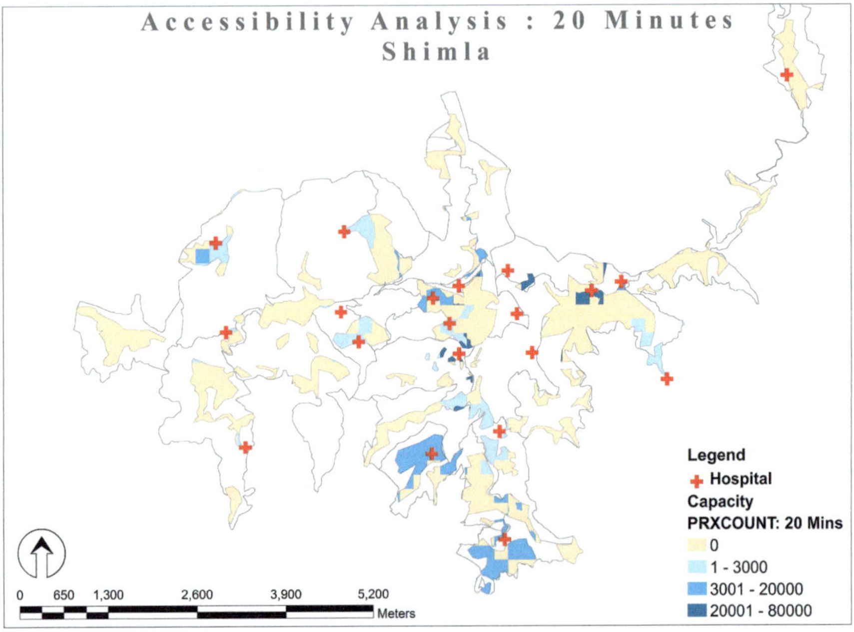

Fig. 1.5 Proximity analysis 20 min in network speed

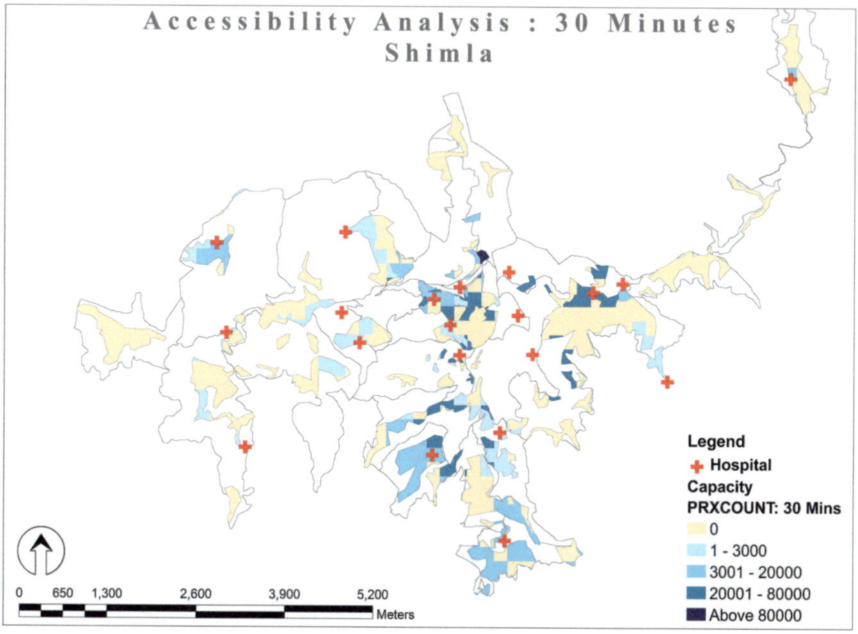

Fig. 1.6 Proximity analysis 30 min in network speed

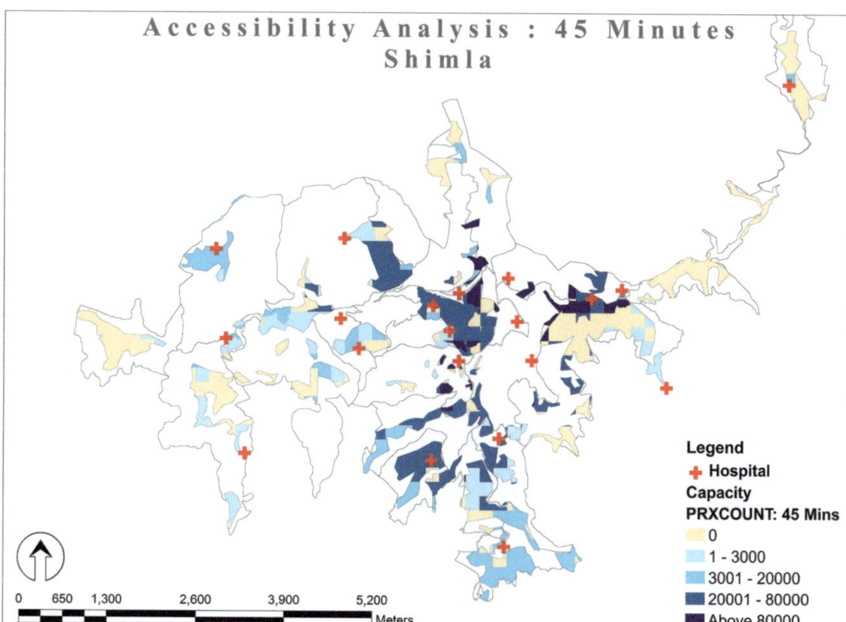

Fig. 1.7 Proximity analysis 45 min in network speed

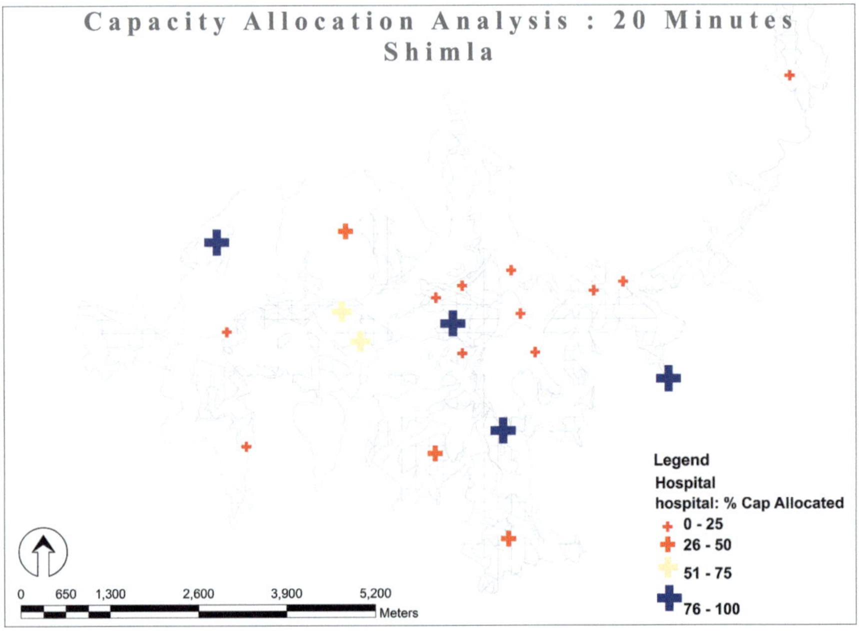

Fig. 1.8 Catchment analysis in 20 min

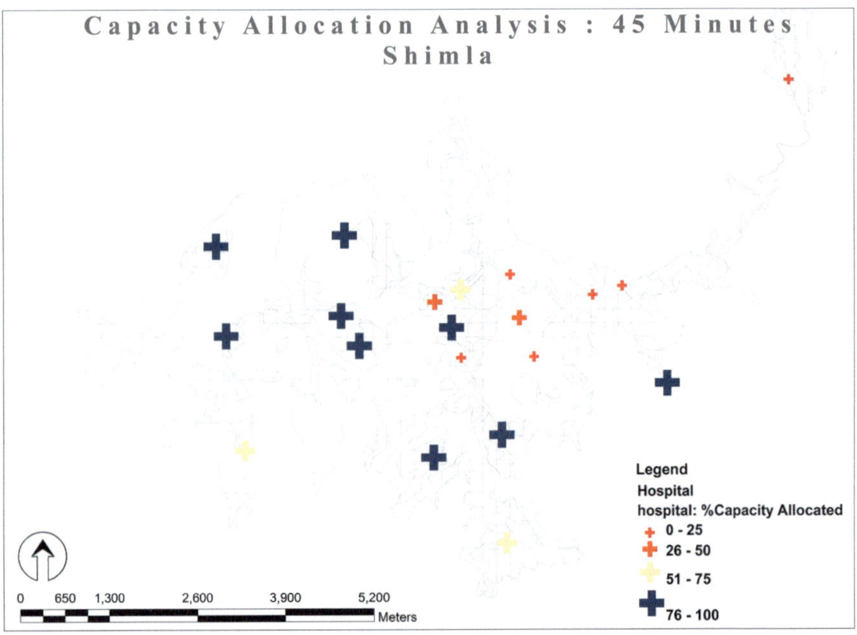

Fig. 1.9 Catchment analysis in 45 min

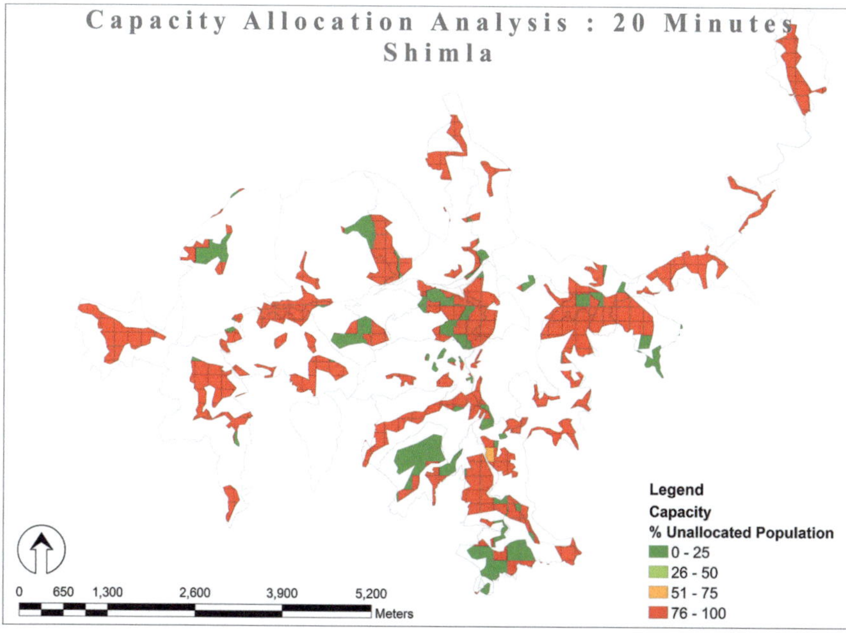

Fig. 1.10 Grid-wise percentage unallocated population in 20 min

1.6.3 Unallocated Areas

Unallocated areas in terms of percentage of population are computed and graded accordingly to its coverage (as Figs. 1.10 and 1.11), red being the least covered area and green being the most covered area.

1.7 Conclusion

The paper intends to present a more methodological way of looking into health infrastructure as a matter of service delivery in terms of social equity from urban planning perspective. This analysis can be developed in detail with a very structured and well-defined data set. And the analysis can be input to multiple policy level interventions, act as an input in development plan-making process, and can be utilised for infrastructure up gradation. Study exhibits that in Shimla has most of residential pockets are inaccessible within a time frame of 20 min. Most areas get accessible when the time limit is extended to 45 min. Looking at the outputs, recommendation will be further diagnose the problem and come up with

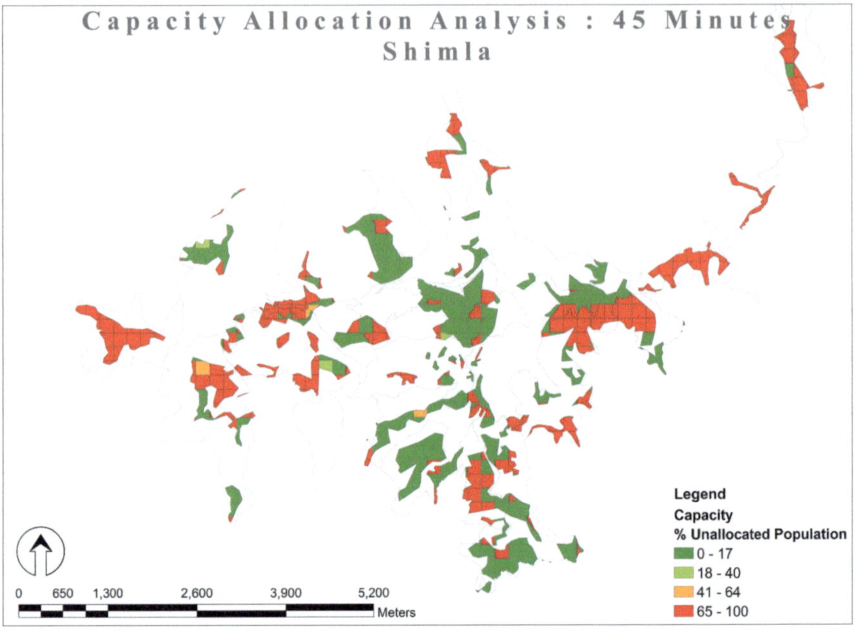

Fig. 1.11 Grid-wise percentage unallocated population in 45 min

measures so that accessibility parameters can be improved drastically atleast in the case of health facilities which is vital for prosperity of community. This study can be applied in the case of urban development, healthcare supply, provisioning, etc.

References

Anvita A, Geetam T (2006) A gendered perspective of the shelter–transport– livelihood link: the case of poor women in Delhi. Transp Rev 26(1):63–80

Carol A (2012) TerraUrban. Retrieved from https://terraurban.wordpress.com/2012/11/06/tangles-of-undelivered-jnnurm-housing-shimla/

Dev Nathan GK (2004) Globalization and indigenous peoples in Asia. Sage, New Delhi

Geurs KT, Eck JR (2001) Accessibility measures: review and applications, evaluation of accessibility impacts of landuse transport scenarios and related social and economic impacts

Mazta S, Thakur A (2012) Primary health care perspectives of village women in Himachal Pradesh. SEARO, WHO

Price M (n.d.) Slopes, sharp turns, and speed. ESRI 8

Sapkota, JB (2015) Human development impacts of access to infrastructure in hills and mountains: evidence from households of three remote villages of Nepal. ISANET. http://web.isanet.org/Web/Conferences/GSCIS%20Singapore%202015/Archive/86c64868-3e9e-40b1-9526-afe8d20c4a08.pdf

Sekhar S (2011) Urban sprawl and other spatial issues in Shimla Himachal Pradesh. ITPI 8:53–56

UNDP and the Planning Commission (2003) District Development Report. SHIMLA, Shimla

Chapter 2
Conservation of Cultural Heritage: The Necessities, Trends, and the Analysis of Current Practices

Ankita Choudhary and Mayank Mausom

Abstract India is a country of multitudes and multivalence, a country so diverse that language changes every quarter of a hundred kilometers. The country is as old as history itself and has some of the oldest cities by the time period over which they have been continuously inhabited like Varanasi, Madurai, Ujjain, etc. (Table 2.1). The country and its innumerable cities and antediluvian settlements have an astounding legacy in terms of heritage and culture.

Conservation and protection of these built as well as concomitant intangible heritage should be an integral part of urban planning, city development, and – importantly – urban design, incorporating architectural conservation, in our urban areas. Old conurbations and heritage built forms are enormously significant as they not only impart variety to our built-up environments, stimulating visual importance, but also give a sense of pride to inhabitants and enthusiasm within the urban settlement.

The current paper tries to help understand the meanings of the key terms in culture and heritage conservation. It explores the necessity and advantages of conservation of cultural heritage in terms of identity, unity, tourism, and economy and the principles that can be adopted for the same. Taking the example of various projects throughout the paper, Pondicherry in specific, the paper tries to develop an understanding as to what should be done and what shall be avoided in conserving the cultural heritage of the place.

Keywords Conservation • Cultural heritage • Tourism • Cultural identity • Culture • Heritage • Tourism

---

A. Choudhary (✉) • M. Mausom
IIT Roorkee, Roorkee, India
e-mail: ankita.choudhary20@gmail.com; mayankmausom@gmail.com

F. Seta et al. (eds.), *Understanding Built Environment*, Springer Transactions in Civil and Environmental Engineering, DOI 10.1007/978-981-10-2138-1_2

Table 2.1 Chronological list of some of the oldest continuously inhabited cities in India

City	Region old	State	Period	Source
Varanasi	Iron Age, North India	UP	ca. 1200 BC	Iron Age Foundation (Painted Grey Ware culture)
Ujjain (as Avanti)	Malwa	MP	800 BC	Rose to prominence in ca. 700 BC as capital of Avanti
Rajagriha (Rajgir)	Magadha	Bihar	600 BC	
Madurai	Pandyan Kingdom	Tamil Nadu	500 BC	There are accounts of Megasthenes (ca. 350–290 BC), a Greek ethnographer in the Hellenistic period
Vaishali	Magadha	Bihar	500 BC	
Patna	Magadha	Bihar	ca. 400 BC	As Pataliputra was founded by Ajatashatru
Thanjavur	Early Chola Kingdom	Tamil Nadu	300 BC	Some scholars believe that the city has been existing since the Sangam period

Source: Overy (1999). The Times History of the World

2.1 Introduction

Cultural heritage is the be quest of physical artifacts as well as the intangible features of a family, society, or settlement which are handed from past generations to the next, preserved in the contemporaneous and imparted for the value of forthcoming generations. Cultural heritage comprises the tangible culture components or the physical manifestations like the buildings, monuments, landscapes, books, works of art, relics, and artifacts and also includes the intangible culture such as folktales, legends, traditions, customs, language, and knowledge and also encompasses the natural heritage. This paper will deal with only the architecture and built-form component of cultural heritage. The Constitution of India under fundamental duties mandates the citizens to value and preserve the rich heritage of our composite culture.

2.2 Cultural Heritage as a Necessity

A city's cultural heritage bears evidence to its ethnic traditions, encourages the development of its modern civilization, as well as makes available monetary benefits such as income through tourism industry. These inherited artifacts are thus customarily looked upon as vital treasures, and conserving these treasures is important so as to sustain the cultural heritage. We, fortunately, have some very old surviving cities which have formed the centers for heritage and culture (shown in Fig. 2.1). If

Fig 2.1 Ancient India map; most of the cities have survived till date (Source: rbi.in)

we do not preserve what is left and conserve what has partially decayed, we might end up losing this wonderful inheritance of ours. Some examples have been shown in Figs. 2.2 and 2.3.

The various components of cultural heritage and their advantages are discussed below.

Fig. 2.2 Neglected site at Arikamedu (100 BC), near Pondicherry (Source: Author)

Fig. 2.3 The 144-year-old Mairie Building which was the town hall in Pondicherry (Source: Author) (INTACH had made a proposal for restoration of the town hall building, and a fund of 7.5 crores was sanctioned by the World Bank)

2.3 Cultural Heritage as a Unifying Factor

As ethnologically diverse societies, like the cities of today, have the propensity for disintegration, and as customary methods of socialization (e.g., education, ethnic policy) have slowly but surely lost efficacy, heritage and its propagation have arguably emerged as a privileged tool for cities to promulgate and celebrate a shared cultural identity with their citizens regarding their city and its past and culture. The sense of pride, thus acquired, goes a long way in unifying the citizens of the urban areas.

2.4 Cultural Heritage as an Identity Creator

While visiting the heritage sites and places of cultural significances, the people understand "who they are and where they have come from." Garcia Canclini[1] suggested that heritage has to be preserved, not discussed or analyzed.

Cultural heritage is significant in creating a regional identity, like we spontaneously visualize the Ghats of Varanasi when we imagine Aarti and the Ganga (shown in Fig. 2.4). The combination of the metaphysical manifestation of the rituals of Aarti and its physical manifestations with the Ghats, steps, and people in the background completes a picture – an identity of that place. This is cherished and forms an image that remains embedded in the consciousness for a long while.

Identity in India has aligned across smaller assemblages grounded on spiritual, cultural, and linguistic identities, thus giving it a much wider variety in identities across its length and breadth (Fig. 2.5).

2.5 Cultural Heritage as Tourism

According to MacCannell, "Tourism is not just an aggregate of merely commercial activities; it is also an ideological framing of history, nature and tradition; a framing that has the power to reshape culture and nature to its own needs." The number of domestic tourist visits in India during 2011 was 850.86 million as compared to 747.70 million in 2010, with a growth rate of 13.8 %.[2] In India, religious heritage places are the foremost tourist attractions that are passionately visited and persistently promoted.

---

[1] Argentine anthropologist.

[2] India Tourism Statistics, 2011.

Fig. 2.4 The excavations at Sarnath; these building footprints are reminiscent of Buddhism and have great connotational value (Source: Author 2013)

2.6 Cultural Heritage: Economic Benefits

The cultural heritage of any city or nation is a great asset in economic terms too – it brings tourism and, along with it, a lot of revenue to the exchequer. Generally, the conservation of heritage buildings and precincts, if carried out in conjugation with a broader tourism plan, prove to be economically sustainable. The latest trend in many Indian cities as inspired from their European counterparts is to integrate business and heritage structures. The historic buildings are being conserved and converted to heritage hotels, emporiums, mini-museums, handicraft stores, heritage restaurants, and much more. The urban precincts, upon conservation can be very tempting addresses for business and markets. Many cities have started having weekly markets in the heritage precincts which again boost local economy. Anyways, in general, even for the building owners, restoration is often cheaper than new construction.[3] The conservation of heritage buildings directly increases the employment in the construction industry – in the restoration works and products

[3] Benefits of Heritage Building Conservation, historic research branch, Manitoba.

Fig. 2.5 Sandhya Mahaarti at the Ghats of Benares; any Banarasi could easily associate with this space and the ritual therein (Source: Author)

manufacturing industry related to it. More skilled laborers get employed and conservationists – architects and builders – get employment as well. The local municipality can also end up gaining in terms of tax, once the old precinct starts thriving in tourism and business. All direct and indirect dependents of tourism industry will get to gain from this long-term planned endeavor.

2.7 Current Practices in Conservation of Cultural Heritage

The premier organization which overlooks all the archaeological research, findings, and conservation of the cultural heritage is the Archaeological Survey of India (ASI). The ASI falls under the Ministry of Culture, Government of India. The major objective of the ASI is the "maintenance of ancient monuments and archaeological sites and remains of national importance".[4]

[4] asihampiminicirecle.in (official website of ASI).

2.8 Issues with Current Practices

2.8.1 *Limited to Listed Monuments*

Only protected and listed monuments are conserved, and funds are allotted for their maintenance and conservation. Cities have significant heritage which is neither listed nor protected and is getting lost by demolitions/changes; only a minor percentage is listed and thus a majority remains neglected, e.g., in Pondicherry the number of heritage buildings as documented by INTACH has reduced to a mere 600 from the erstwhile 2000 in the year 1996. At this pace, we might end up losing much of the tangible heritage we so proudly claim. ASI is not as proactive as it was envisaged to be.

2.8.2 *Barricading and Charging Fees for Entry to ASI-Protected Sites*

It might appear to be a nominal fees of rupees ten, but it eliminates the possibility of entry of poor people, and we have a substantially large poverty-ridden population. Don't the poor have the right to cherish and enjoy public spaces and their heritage monuments? This also reduces the frequency of possible visits by local populations and thus eliminates the prospective of conception of public spaces in the purlieu of these built forms.

2.8.3 *Pitiable Quality of Conservation Work*

Due to the lack of skilled craftsman and labor or lack of will, the restorations of ASI are often of awfully inferior quality. A glaring example can be seen in Lucknow, wherein the famous Bara Imambara's Daulat Khana (entry) was conserved by ASI as seen in Fig. 2.6. The restored Daulat Khana building and the Baoli in Lucknow, both disinherited of their originality and multifariousness during the process of restoration; it now stands in obdurate contrast with the rest of the iconic Imambara complex.

2.9 Heritage Legislations in India

Some legislations related to heritage monuments and their conservation are listed below.

Fig. 2.6 The restored Daulat Khana building and the Baoli both disinherited of their originality and multifariousness during the process of restoration; it now stands in obdurate contrast with the rest of the iconic Imambara complex

2.9.1 By the Central Government

The Ancient Monuments and Archaeological Sites and Remains Rules 1959, is only restricted to the "protected" monuments in the country. This was an amendment to an earlier Act of 1904 and 1951 framed by the Central Governments.

2.9.2 State Town and Country-Planning Acts

Maharashtra has been a pioneer of heritage legislation: Maharashtra Regional and Town Planning Act, 1966, as modified up to 1988; Local Urban Areas, Metropolitan Areas, Regional Level Amendment in 1995; BMRDA Act, 1975; MHADA Act, 1976; and Maharashtra (Urban Areas) Preservation of Trees Act, 1975. This triggered Model Regulations circulation by MOEF to all state governments and union territories on 28 December 1995 which prompted many states like Andhra Pradesh, Punjab, Delhi, Gujarat, Goa, Bihar, UP, MP, Rajasthan, etc. to further strengthen their existing regulations.

2.9.3 Local Development Control Rules/Byelaws

Not many cities are doing anything substantial at the local level except perhaps Pondicherry and Delhi, which again are states. Delhi has an Urban Arts Commission and being the national capital has certain privileges by ASI and some NGOs as well.[5] Though there has been some progress in the legislation part, the ground

[5] INTACH and Aga Khan Trust.

implementations seem to be missing in most states; conservation has to be on field, outside of the papers.

2.10 Suggested Conservation Principles and Objectives

Conservation ought to be considerably more than just preserving a building frontage or façade or the superficial elemental superimposition on a building or barricading and claiming that it is protected. It is of paramount significance that we retain the innate essence and original ambience of these historic buildings to the extent conceivable. The conservation of these constructions and heritage precincts in the urban areas shall be exemplars of our opulent architectural, remarkable, historic, and rich cultural inheritance. Conserving the historic structures also enhances the idiosyncratic character and unique identity of the cities. This necessitates an intrinsic sense appreciation, empathetic knowledge, and deep understanding of the architectural structure of the buildings, good management, and practices in conserving buildings.

2.10.1 The "3R" Principle of Conservation[6]

Maximum ***Retention*** *, Sensitive* ***Restoration*** *, and Careful* ***Repair***

This principle has been successfully implemented by the Urban Development Authority of the Government of Singapore and forms the core guiding principle behind all the architectural conservation work carried out in Singapore (Fig. 2.7). The key is to retain the originality and essence of the architecture of the historic buildings to the maximum possible extent without reconstructing the whole of the historic built structure. It is very essential to recreate the buildings in such a way so as to recreate the original ambience of the buildings. It is also very important to do intensive research and documentation before proposing any sort of conservation work, preferably by hiring professionals. The quality of the conservation work must not be compromised in any case, and carefully executed accurate repairs should be applied. The conservation work has to go beyond just retaining the original façade of the historic building. Smart materials and latest technologies like 3D printing shall be applied, if necessary, to attain the aforementioned objectives.

[6] Urban Development Authority, Government of Singapore.

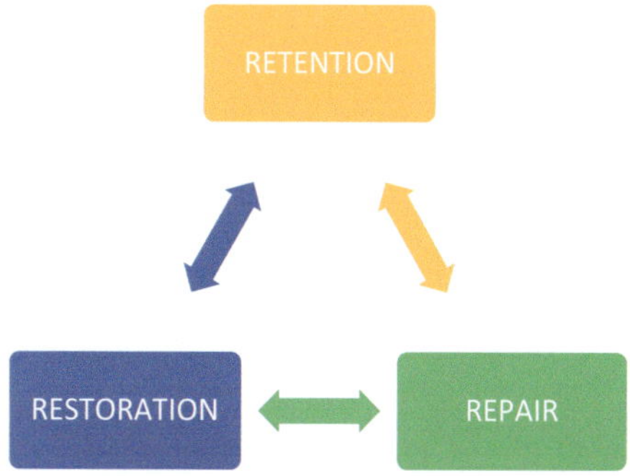

Fig. 2.7 The 3R principle of heritage conservation

Fig. 2.8 The image ensemble of the pre-restoration and post-restoration phases of Nagaur Fort (Source: Archnet)

2.11 Exemplars of Successful Conservations

Some good examples of conservations and revitalizations can be used as models and ideal practices.

2.11.1 Restoration of Nagaur Fort in Rajasthan by the Mehrangarh Museum Trust (Shown in Fig. 2.8)

The Getty-supported conservation work was recognized with a UNESCO Award for Excellence in Cultural Heritage Conservation.

Fig. 2.9 The picture ensemble showing the pre-conservation and post-conservation phases (Source: Aga Khan Trust for Culture)

2.11.2 Restoration of Humayun's Tomb and Gardens by the Aga Khan Foundation

Another excellent example of heritage building restoration is the revival of Humayun's tomb and the Charbagh amidst which it stands as shown in Fig. 2.9. This was done by the Aga Khan Foundation.

2.11.3 Listing of Buildings and Restoration of Vysial Street, Pondicherry, by INTACH and Pondicherry Government

Streetscapes of the Tamil Precinct. Tamil streets are mainly characterized by "thalvarams" – a social extension of the house – and "thinnais." INTACH came forth and restored some of the buildings in Vysial Street (Figs. 2.10 and 2.11).

2.12 Conclusion

There are multiple reasons as to why the traditional settlements and the old parts of cities shall be conserved. The old urban precincts have cultural and social values associated with them apart from holding economic, historical, and also esthetical importance. People are generally very proud of their old cities and have certain associational attachment to their heritage. The tangible heritage, the buildings, and old precincts contribute to the identity and perceptibility of cities. They evoke emotion, nostalgia, thought, and pride by showcasing the past and the background. Heritage is not only about the past. It is about the present generation who continues to cherish and to learn about the effervescent and glorious history, culture, and past civilization. It is about cultural traditions, places, and values that people proudly

Fig. 2.10 Vysial Street, Pondicherry (Source: INTACH)

Fig 2.11 The map of heritage buildings (INTACH)

preserve (Collins 1983). The conservation of cultural heritage is a precursor to urban regeneration; if the old parts are taken care of, the city in totality will look good and feel good to live in; the government organizations have to be more proactive in this regard; and creation of conservation cells at a local level, involvement of more NGOs, public participations, and PPP models will certainly help in achieving that. Urban heritage goes beyond just buildings; it is about the way a settlement evolves in relationship to its natural surroundings; it's about the streets and public spaces as well as the utilities. Conservation of heritage concerns urban planning and development and the need to respect and construct on indigenous, natural, cultural, social, and economic resources.

References

Collins R (1983) Tourism and heritage conservation-The Pacific Experience. Herit Aust 2(2):58–59

Overy RJ (1999) The times history of the world: new edition. Times Books/Harper-Collins, London

Chapter 3
Climate Responsiveness of Wada Architecture

G.M. Alapure, Abraham George, and S.P. Bhattacharya

Abstract India has vast diversity in climatic conditions, traditional wisdom, and cultural notions which have influenced its vernacular-traditional architecture. The traditional strategies have evolved over time, using locally available materials, depicting a technological peak of its period. Climate is one of the significant factors which influence orientation, form, organization, and pattern of built forms and the use of materials. The research attempts to identify passive-cooling strategies employed in Wada, a vernacular-traditional built form, in the moderate climatic zone of Western Maharashtra. The research study comprises a comprehensive literature review, definition, selection of case studies, and field work. A survey is carried out by recording on-site observations, following an instrumentation setup and measurement procedure. The paper documents the appropriate principles employed in vernacular-traditional built forms that are sustainable and are expected to be of relevance to architects and built form designers.

Keywords Wada • Vernacular architecture • Traditional architecture • Sustainable built form • Thermal performance

3.1 Introduction

Architecture is a physical manifestation of the needs and aspirations of a society and is determined by the environmental, socio-cultural, and political climate of a place. It includes not only monumental and professionally designed monuments but also quaint residential structures built by craftsmen and lay-people. It includes the interrelationships of the built and open spaces (Cooper and Dawson 1998). Wadas are popular as a form of residential architecture of *Marathas* under the patronage of *Peshwas*, in India.

These were built all over the Maharashtra region to serve the distinctive needs of big joint families. Interestingly, though Wadas housed the *Brahmin aristocracy*, the

G.M. Alapure (✉) • A. George • S.P. Bhattacharya
Department of Architecture & Regional Planning, Indian Institute of Technology Kharagpur, Kharagpur, West Bengal, India
e-mail: gmalapure@rediffmail.com; abrahamiitkgp@gmail.com; spb@arp.iitkgp.ernet.in

F. Seta et al. (eds.), *Understanding Built Environment*, Springer Transactions in Civil and Environmental Engineering, DOI 10.1007/978-981-10-2138-1_3

built form shows a significant *Mughal influence* in their design through the integration of courtyards with fountains, pools, terraces, arches, lattice windows, and balconies. Symmetry is an indispensable aspect of these houses. Wadas rise to two to three floors. Many urban and rural Wadas have smaller courtyards due to shortage of space. However, most of these houses have a courtyard in the front to receive visitors and conduct business and administrative activities of the head of the family. The urban form of the settlement is a maze of a two or three-storied structure having internal courtyards, placed adjacent to the road network with a little community space. Wada is not synonymous with "plot of land." However, by the eighteenth century, it became a tradition to place a house, large or small, under the nomenclature "Wada" (Mate 1959).

3.2 Methodology

The research is conducted in a moderate climatic zone of Western Maharashtra. The investigation consists of a comprehensive review of relevant literature, definitions, selection of the case studies, and field works. This is a combination of extensive theoretical research studies and painstaking fieldwork in which the following objective is achieved:

- To evaluate existing passive-cooling strategies employed in Wadas in terms of their thermal performance.

To achieve the objective as mentioned above, the relevant data are collected from existing literature. Further, in situ measurements are recorded by using a *vane anemometer*, and personal experiences are obtained in response to the open-ended questionnaire. Case study is chosen from a range of vernacular-traditional, passive-cooled *Wadas* on its thermal performance. Experimental work recorded, dry-bulb temperature (DBT) which is fairly representing human comfort along with relative humidity (RH). Observations of DBT and RH were taken at hourly intervals by using a *vane anemometer (ISO 9001; certified model, AH-4223; accuracy ± 1.5 °C)*. Indoor and outdoor climatic measurements are recorded at 15 cm above the floor level, 120 cm high from floor level that defines activity zone, and 20 cm below the ceiling, to understand the effective temperature at ceiling level (Koenigsberger et al. 1975). Climatic elements' measurement recording is carried out from 6:00 am to 18:00 pm on December 21 and 22, 2013, and March 21 and 22, 2014.

During these equinoxes' dates, the rays of the sun shine directly on the equator. This occurs in summer days on March 20 and September 22 and in winter days on December 21, in one of the two solstices, when the rays of the sun directly strikes one of the two tropical latitude lines.

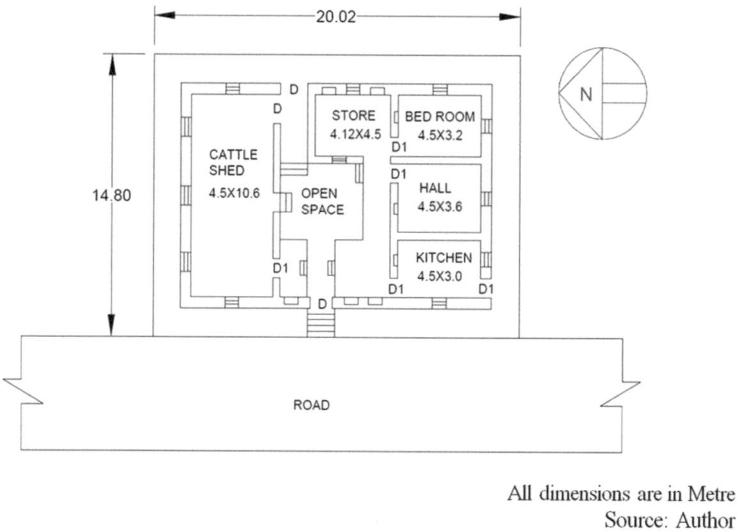

Fig. 3.1 Plan of Raghoba Salunkhe, case study Wada (All dimensions are in meter)

3.3 Case Study

Raghoba Salunkhe Wada, House No. 38, a naturally ventilated built form in the village of Basarapur, Bhor Taluka, of Pune District in Western Maharashtra, India, with a latitude of 18.17° N and 73.85° E, is investigated. The investigation is carried out on December 21 and 22, 2013, and March 21 and 22, 2014. The measurements of climatic elements were not recorded on June 21 and 22, 2015, due to cloudy sky and heavy rain. The case study, Wada, is a single-storied structure built in 1905 almost 109 years ago. The Wada is surrounded and overlooked by a veranda, central open courtyard, reception hall, and other rooms as shown in Figs. 3.1 and 3.2. The material used for walling is dressed igneous and sedimentary rock. The igneous rock is dark colored and fine grained, composed generally of plagioclase and pyroxene minerals. The igneous rock, basalt, is locally and easily available, cheap, durable, and low-embodied energy-building material which is used in Wada construction. The sedimentary rocks like sandstone and limestone are often used in external walls. The sedimentary rocks are well suited to carving into blocks of any shape. Shale can be broken up and crushed to make desired size blocks. It can be easily used with lime mortar. Doors and windows are made up of locally available wood. The hall, kitchen, and other rooms are pitched roofed with terra-cotta tiles and having lattice windows or *zarrokas*. Bedroom and entrance passage roofs are made of flat mud roof, locally called *malwad* or *tulai*, supported on wooden beams and purlins.

Fig. 3.2 Case study Wada in Basarapur, Bhor Taluka, of Pune District in Maharashtra, India

3.4 Results and Discussion

The Graph 3.1 shows that the DBT variations during 10:00 am to 17:30 pm for indoor temperature are less than outdoor temperature; whereas indoor temperature is varying at 5 °C (min 22 °C to max 27 °C), variations in outdoor temperature are at 11 °C (min 22 °C to max 33 °C) on December 21 and 22 of 2013 in daytime. Indoor temperature is warmer during morning hours which is comforting in cold winter days. Indoor warm period is prolonged by 7 hours during daytime in winter season.

The Graph 3.2 shows that in the summer, from 6:00 am to 17:30 pm, the indoor temperature is relatively cooler in comparison to outdoor temperature. Temperature variations between indoor and outdoor temperature during the daytime are quite high, 6–8 °C, which indicates the passive-cooling efficiency of the Wada.

Human response to the thermal environment does not depend on air temperature alone. It has been established beyond doubt that air temperature, humidity, radiation, and air movement, all produce thermal effects and must be considered simultaneously if human responses are to be predicted (Koenigsberger et al., 1975).

In the Graph 3.3, indoor and outdoor DBT and RH values are recorded at specified locations (mostly at the center of each room and at the midpoint of the road). The winter and summer climatic elements describe the comfort zone (20–30 °C and RH 30–65 %, Bio-climatic Chart by Olgyay), where there is no perceivable wind speed which is greater than 0.25 m/s and the RH near 80 %. It has been substantiated from the Graph 3.3 that most of climatic element measurements of DBT vs. RH are

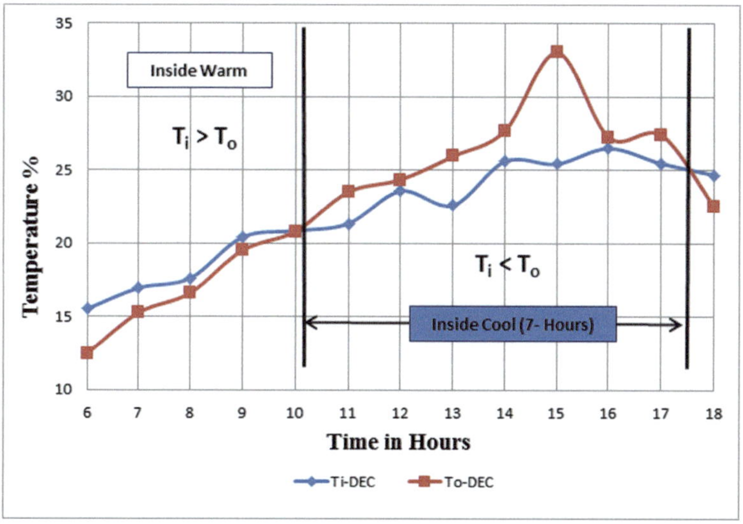

Graph 3.1 Time (in hours) vs. indoor and outdoor DBT (in 0 °C) recorded on December 21 and 22, 2013

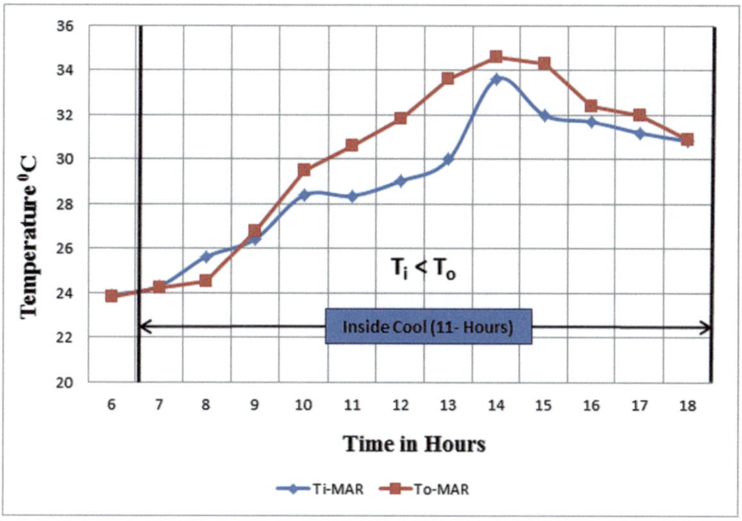

Graph 3.2 Time (in hours) vs. indoor and outdoor DBT (in 0 °C) recorded on March 21 and 22, 2014

falling in the comfort zone, and moderate, and cold humid zone. Very few measurements are falling in hot-dry and cold-humid zone, but no measurement is falling in hot-humid zone and cold-dry zone. The house uses an attic space not only as a storage space but also as a barrier between the hot summer sun and living space. Attic space helps to reduce the temperature of the ceiling of the living zone in summer. It

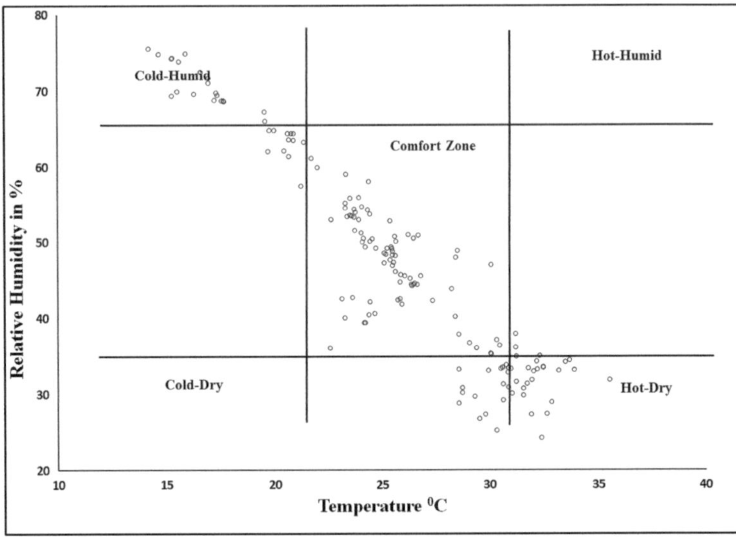

Graph 3.3 Bioclimatic chart showing DBT (in 0 °C) vs. RH (in %) recorded on March 21 and 22, 2014

Table 3.1 Passive-cooling strategies employed in Wadas

Sl. No.	Wada inbuilt components	Climatic impact
1.	Courtyard with verandah	Courtyard with verandah maintains a pool of cool air. Shading walls from radiant heat. Acts as microclimate modifiers. Verandah acts as a buffer for light and ventilation within the room. Acts as sun face to wings.
2.	Fountains and pools	Fountains and pools add to the RH, brings in evaporative cooling to adjacent rooms when wind is drawn over it and into rooms.
3.	Terraces; shading devices	Prevents from radiant heat exposure.
4.	Lattice windows; arches	Smaller-sized lattice windows offer ventilation. Prevents radiant heat and sky components.
5.	Attic spaces	Offer thermal resistance with ventilated space to remove hot air. Attics reduce radiation intensity on horizontal surface. Serves to lower MRT inside. Used for storage.
6.	Locally produced high thermal mass for roof and walls	Offer longer time lag and prevents the heat build up indoors with thermal reversal and acts as sound and dust insulator
7.	Water sprinkling in summer evening	Water sprinkling adds to the RH, brings in evaporative cooling to adjacent rooms to the courtyard. Prevents dust and suspended particles in indoor air.
8.	Symmetry in plan	In a square building, each side will receive equal radiations.

is observed that as a daytime activity water is sprinkled in the courtyard in summer evening. It adds to the moisture content, also, thanks to the evaporative cooling effect to adjacent rooms in the courtyard. A few key and passive-cooling strategies employed in Wadas are listed in Table 3.1.

3.5 Conclusions

It could be concluded that the "Wada" built forms offer positive benefits in regard to thermal performance. Attic space helps to reduce the temperature of the ceiling of the living zone in summer. Sprinkled water in the courtyard in summer evening adds to the RH and brings in evaporative cooling to adjacent rooms in the courtyard. The design elements of *Wada* incorporate passive comfort strategies evolved over successive generations. The inclusion of courtyard with verandah, fountains, pools, terraces, attic spaces, lattice windows, arches and use of low-heat conductive, high thermal mass lead to passive-cooling strategies employed in Wada in terms of their high thermal performance. Comparatively larger thermal mass of Wada helps to create stable conditions inside the built form. Day and night time temperatures are thus lowered using the traditional passive means in Wada. These vernacular-traditional built forms are climate responsive with inbuilt passive-cooling techniques. Thus, it may be seen that vernacular-traditional architecture works in the lines and principles of passive means to achieve sustainable built form and stays relevant in contemporary times too.

References

Koenigsberger OH, Ingersoll TG, Alan M, Szokolay SV (1975). In:Manual of tropical housing and building climatic design. Orient Longman Pvt.Ltd, Hyderabad, pp 31–51.
Liay C, Barry D (1998). Traditional buildings in architecture. Thames & Hudson, London.
Mate MS (1959). Maratha architecture. University of Pune, Poona, p 22.

Chapter 4
Investigating the Architectural Manifestations of Path and Place in Sacred Sikh Architecture

Ripu Daman Singh, Jatinder Kaur, and Prabhjot Kaur

Abstract The paradigm of sacred architecture, since early times, has been guided by elements of path and place making. Previous researches have shown that there are specific architectural elements which correspond to the path and place making in sacred spaces. This paper aims to analyze the path- and place-making elements proposed in previous researches and develop a comprehensive list of the same for sacred Sikh architecture, i.e., Gurdwaras. The list, thus developed, shall help to rediscover the underlying pattern of path and place making which contribute toward the sacredness in Sikh architecture. The study also helps to rediscover the architectural elements responsible for the uniqueness of these memorable spaces and which are reminders of events and experiences that took place in them. Few prominent historical Indian Gurdwaras have been included in the study for identifying the architectural elements signifying the presence of path and place making. The study concludes with the summary of architectural manifestations of sacred Sikh architecture which correspond to the categories of path and place.

Keywords Sacred Sikh architecture • Gurdwaras • Path and place • Architectural manifestations

---

R.D. Singh (✉) • J. Kaur
Department of Architecture, GZSCCET, Bathinda, India
e-mail: ripu_jatinder@yahoo.co.in; ar.jatinder@gmail.com

P. Kaur
Department of Built Environment, PIT, Mohali, Punjab, India
e-mail: pkaurdap@gmail.com

F. Seta et al. (eds.), *Understanding Built Environment*, Springer Transactions in Civil and Environmental Engineering, DOI 10.1007/978-981-10-2138-1_4

4.1 Introduction

A sacred space is set apart from the usual context of life, and its practical use of worshipping is not its dominant feature (Osbon 1991). The paradigm of sacred architecture is strongly influenced by the characteristics of path and place. The presence of path- and place-making elements helps in accentuating the sacred character of a place (Tabb 2006). The modern building methods have overtaken the traditional character of place making, which was influenced by wisdom and craft (Lyndon and Moore 1994). In today's context, the timeless principles of design that guided the sacred path and place making are overlooked or forgotten. The need is, therefore, to reconnect with these principles of design in order to provide inspiration and renewed interest in principles of sacred architecture (Timotio 2008). This study shall help to identify the various path- and place-making elements which are reminders of events and experiences that took place in them (Lyndon and Moore 1994). The spatial organization in sacred spaces depends upon the presence and the dynamic relationship of path- and place-making characteristics (Barrie 1996). Therefore, there is an urgent need to reconnect with the principles of sacred architecture which guided their design since the early beginnings (Timotio 2008).

4.2 Significance of Path and Place in Sacred Architecture

Path and place are identified as important characteristics to understand the complexity in sacred architecture (Mazumdar Shampa and Mazumdar Sanjoy 2004). It is speculated that the spiritual energy imbibed in the characteristics of path- and place-making elements in sacred architecture contributes to sacredness (Tabb 2006). The path, toward and through a sacred space, differentiates it from other spaces, making it very important in sacred architecture. The space is transformed into a place by walking and movement through the path (Fran 2012). Places are distinct spaces that one can remember, making them a part of one's life, reach on their feet, and fill with their presence. These are guided by the geometrical principles based on point, line, plane, and solid depicted through various elements of built form (Lyndon and Moore 1994).

4.3 Sacred Sikh Architecture: Gurdwaras

Sikhism as a religion is comparatively of recent origin when viewed in context of other age-old religions in India. Started by Guru Nanak Dev Ji, the first Sikh Guru, Sikhism evolved as a religion at the time of the fifth Sikh Guru, *Guru Arjan Dev Ji* (Arshi 1986). The Sikh Shrine, earlier known as *Dharamsal*, came to be known as *Gurdwara* with the enshrinement of *Guru Granth Sahib* in the Golden Temple, Amritsar, in 1604. The term *Gurdwara* is suggestive of a structure having a door and a seat for placement of *Sri Guru Granth Sahib* which thus contains an architectural connotation (Singh 2007).

The sacred Sikh architecture tends to develop into a complex of several buildings serving different functions, expressing the three commandments of Sikh faith, as per *Guru Nank Dev Ji* – worshipping (*Naam Japna*), working (*Kirt Karna*), and sharing (*Vand Chhakna*). *Gurdwaras* are open to one and all without any distinction of caste or religion, depicted in plan by entrances in all four directions. The external distinguishing mark in a *Gurdwara* is *Nishan Sahib* – the Sikh flag of saffron or blue color. *Gurdwara* complex is entered through an entrance portal called *Deorhi* – depicted as an impressive gateway from where one gets a glimpse of the sanctum sanctorum (Bhatti 1987).

4.4 Literature Review

Path and place are identified as important characteristics to understand the complexity in sacred architecture (Mazumdar Shampa and Mazumdar Sanjoy 2004). Many authors have researched and developed fundamental characteristics associated with path and place making.

In 1985, Michael Brill proposed a list of 14 place-making patterns, which are theorized to be present at sacred spaces. The description of these design characteristics was given in the address titles, "Using the Place-creation myth to develop design guidelines for sacred space," at the annual conference of the Council of Educators in Landscape Architecture at the University of Illinois in Urbana, IL, USA. In 1986, the Encyclopedia of Religion edited by Mircea Eliade, the architectural typology of religious buildings, was given. This typology was based on the categories of path and place so as it makes the spaces more identifiable. In 1994, Donlyn Lyndon and Charles Moore prepared a list of specific themes and compositions in their book titled "Chambers for a Memory Palace," which highlighted the elements that make the places memorable. In 1996, Philip Tabb, prepared a series of 15 place-making patterns in his work titled "Sacred place: the presence of Archetypal patterns in place creation" making elements (Timotio 2008).

4.4.1 Summarizing Path- and Place-Making Characteristics

The review of previous researches has been listed in Table 4.1, summarizing specific path- and place-making patterns which are theorized to be present in sacred spaces. The common characteristics have been consolidated so as to prepare a comprehensive list. The comprehensive list, thus developed, shall be used for the analysis of sacred Sikh architecture, i.e., *Gurdwaras* w.r.t. path- and place-making characteristics.

Table 4.1 Comparing and summarizing path- and place-making characteristics

S. No	Design characteristics proposed by Michael Brill (1985)	Architectural typology of religious buildings based on categories of path and place, as per Encyclopedia of Religion (1986)	Elements that make places memorable as per Lyndon and Moore (1994)	Place-making patterns proposed by Philip Tabb (2006)	Summary of characteristics to form a *COMPREHENSIVE LIST*
1.	Making a location and center	Concentrated form sharp terminal	Axes that reach	Center	Central axial focal point
2.	Making orientation and direction	Directionality		Direction	Directionality highlighted by axis
3.	Spatial order	Readily comprehensible shape	Order that comes and goes	Spatial order	Spatial order based on symmetry and memorable forms
			Space that leaks up into light	Numeric order	
			Shapes that remind	–	
4.	Celestial order	–	–	Celestial order	–
5.	Differentiating boundaries	End-from-end distinction	Walls that layer	–	Walled enclosures signifying hierarchy
6.	Reaching upward	–	Roofs that encompass, canopies that center	Ascent	Vertical elements emphasized by roofs forms
7.	Triumph over underworld	Capable of being experienced as an inside in contrast to a surrounding exterior	Platforms that separate, stairs that climb and pause	Descent	Descent leading to sunken spaces
8.	Bounding	Strong edges	Borders that control	Bounding	Boundary defining edges
9.	Passage	Continuity	Paths that wander, portals that bespeak	Passage	Passages through portals signifying continuity and distinction
10.	Ordered views	–	Openings that frame	Ordered nature	Framed openings restricting views
11.	Light	–	Light that plays	–	Light admitted from above

(continued)

Table 4.1 (continued)

S. No	Design characteristics proposed by Michael Brill (1985)	Architectural typology of religious buildings based on categories of path and place, as per Encyclopedia of Religion (1986)	Elements that make places memorable as per Lyndon and Moore (1994)	Place-making patterns proposed by Philip Tabb (2006)	Summary of characteristics to form a *COMPREHENSIVE LIST*
12.	Materials for making	–	Ornament that transmits, transforms, and encodes	–	Ornamented unique and rare materials
13.	Nature in our places	–	Water that pools and connects	Ordered nature	Presence of natural elements
14.	Finishing a place	–	–	Ceremonial order	Ceremonial path
15.	–	Recognizable landmarks	Markers that command	–	Landmarks and markers
16.	–	Limited in size	Rooms that define	–	Modest size

4.5 Architectural Manifestations of Path- and Place-Making Characteristics in Sacred Sikh Architecture

The summary of path- and place-making characteristics has been prepared after comparing the elements proposed by previous researches. The same is presented in Table 4.2, separately for path-making characteristics and place-making characteristics in order to analyze their architectural manifestation in *Gurdwaras*. The analysis, thus presented, shall help to nurture spark in the modern designers and shall provide suggestions for building that make them capable of conveying ideas, in turn making them valuable and being experienced as sacred.

These path- and place-making characteristics, in the context of the present study, are the spatial or physical attributes that aid in path and place making. These characteristics are described with reference to Sikh shrines, where the elements (building parts) present in *Gurdwaras* and the actions (which take place in them) explain how these elements transform that place into a sacred place. The study shall help in gaining an insight into the traditional wisdom of path and place making, which makes the sacred spaces more memorable and distinct.

Table 4.2 Analysis of path- and place-making characteristics through their architectural manifestations in sacred Sikh architecture

S. No	Summary of the characteristics to form a COMPREHENSIVE LIST	Related to path or place	Presence in sacred Sikh architecture	
1.	Directionality highlighted by axis	Path	The primary factor that gives directionality to a path is circumambulation, i.e., desire to encircle a sacred center (Brill 1985). In the *Golden Temple, Amritsar*, the circumambulatory path (parkarma) gives direction and subtly leads one to experience the surroundings yet directs one to the main shrine	
2.	Passages through portals signifying continuity and distinction	Path	The path forms a continuity and means of communication between the sacred and the profane world (Brill 1985). The Path at *Takht Sri Hazur Sahib, Nanded*, symbolizes continuity as it encircles the shrine and guides the worshippers toward the place of worship	
3.	Descent leading to sunken spaces	Path	The *Golden Temple, Amritsar*, is positioned in the center of the sacred pool (sarovar) which itself is positioned much below the surrounding ground. The terminal is highlighted as one descends the steps down the Darshani Deorhi, and the visual connect is always maintained	
4.	Ceremonial path	Path	The path from the threshold of the shrine to the focal point helps in framing the ritualistic movement within a sacred space formed by the strong edges of the path (Brill 1985). In the *Takht Sri Hazur Sahib, Nanded*, the path from the threshold, which frames the view of the shrine, has strong edges in the form of a pedestrian path	

(continued)

Table 4.2 (continued)

S. No	Summary of the characteristics to form a COMPREHENSIVE LIST	Related to path or place	Presence in sacred Sikh architecture	
5.	Landmarks and markers	Path and place	The geometry of the path is affected by the set of recognizable points along it (Brill 1985). The pathway in the *Golden temple, Amritsar*, is dotted with recognizable landmarks such as Langar Hall, Parshad Ghar, Burj, and Bungas, making it all the more legible and is capable of inspiring and affecting even the casual tourist	
6.	Spatial order based on symmetry and memorable forms	Place	Sacred spaces embody the principles of visual geometry through a fixed relationship with the center. The shape must be comprehensible, i.e., clearly understood (Tabb 2006). The *Takht Sri Hazur Sahib, Nanded*, has a simple comprehensible geometrical form in plan and cubical mass dominated by onion domes, turrets, and kiosks	
7.	Boundary defining edges	Place	It signifies a fixed relationship of the center with the comprehensible surrounding edge comprising of unity in the center where all parts are unified into one and diversity at the periphery (Tabb 2006). It symbolizes a boundary or interface between the sacred area and the outside profane world (Brill 1985). The *Takht Sri Hazur Sahib, Nanded*, compound is approached through the entrance gateway, Darshani Deorhi, and the colonnades on all sides provide a sense of enclosure	

(continued)

Table 4.2 (continued)

S. No	Summary of the characteristics to form a COMPREHENSIVE LIST	Related to path or place	Presence in sacred Sikh architecture	
8.	Vertical elements emphasized by roof forms	Place	The vertical elements define the place of sacred space in the vicinity and serve as a guide. It also establishes the geometry in the plan, thus bringing an order. In sacred Sikh architecture, the special silhouettes of domes, chhatris, and towers locate the *Gurdwara Shaheedan*, *Amritsar*, in one's memories and remind of the rich culture	
9.	Central axial focal point	Place	Sacred place has a fixed location and an articulated center, and boundaries related with the four directions are fixed, clear, and distinct (Brill 1985). The *Golden Temple*, *Amritsar*, is like a lotus in the pool of nectar (sarovar) and serves as the focus of the entire complex and is the hub of activity being axially placed in line with the causeway. This makes the focus for gathering even more pronounced	
10.	Framed openings restricting views	Place	Doorways and windows enhance people's expectations of places that lie beyond, drawing views through them. Openings and fenestrations are like eyes, making a rhythm across the facades. The cusped foliated arches of *Takht Sri Hazur Sahib*, *Nanded*, provided at strategic locations and framed windows provide breathtaking views of the sacred space	

(continued)

Table 4.2 (continued)

S. No	Summary of the characteristics to form a COMPREHENSIVE LIST	Related to path or place	Presence in sacred Sikh architecture	
11.	Light admitted from above	Place	Spaces are appreciated more if the contained space escapes around corners or lifts up into clerestory light. The central hall in the *Golden Temple*, *Amritsar*, has a double height roof surrounded by a gallery above admitting ample light below	
12.	Ornamented unique and rare materials	Place	The ornament and use of unique and rich materials identify the culture of a religion. It also depicts the shrine's position in the community and gains attention. The gold plating, inlay work, and surface decoration are deeply embedded in *Takht Sri Hazur Sahib*, *Nanded*, which initiate a series of associations with sacred places	
13.	Presence of natural elements	Place	The natural elements, i.e., trees, water pools, etc., have an extraordinary quality of cooling presence that connects the infinite God with the intimate worshipper. The presence of trees with historical significance (Beris) and sacred-bathing water pools (sarovars) signifies the presence and utmost importance given to nature in the *Golden Temple*, *Amritsar*	
14.	Modest size	Place	The built form and details are based on anthropomorphic attributes, i.e., measurements of the human body expressed in terms of scale (Tabb 2006). The main shrine of the *Golden Temple*, *Amritsar*, and most of the other historical Gurdwaras are modest structures with limited seating	

4.6 Conclusion

The presence of the elements of path and place has a great impact in contributing to the sacredness of a place. The elements of path and place making have been integrated in the sacred Sikh architecture as a meaningful organic experience. The architectural markers of axis, pathways, descent, gateways, and landmarks contribute toward path making, and geometry, symmetry, bounding, vertical elements, focal point, framing of views, filtered light, modest size, ornamentation, and natural elements highlight the place-making elements. Paths and places are exceptionally well defined, highlighting the entrance descent or ascent to the compound, a reflection of the shrine in the sarovar, a causeway leading one to the main shrine positioned in the center of the complex or the pool in a perfect transition to a sacred world, thus setting the stage for a religious experience. This study has shown that there are identifiable architectural manifestations in Gurdwaras related to the dual typology based on the categories of path and place making, which contribute toward the Gurdwaras being experienced as sacred.

References

Arshi PS (1986) Sikh architecture in Punjab. Intellectual Publishing House, New Delhi
Bhatti SS (1987) An outline of Sikh architecture. Punjab University, Chandigarh
Brill M (1985) Using the place-creation myth to develop design guidelines for sacred space. Proceedings of the council of educators in landscape architecture's annual conference, University of Illinois, Urbana
Eliade M (1986) Encyclopaedia of religion. Thomson Gale, Farmington Hills
Fran L (2012) The architecture of pilgrimage: sacred site and sacred self in the quest of meaning. M. Arch. thesis, Victoria University, Wellington
Lyndon D, Moore CW (1994) Chambers for a memory palace. The MIT Press, Massachussets
Mazumdar Shampa, Mazumdar Sanjoy (2004) Religion and place attachment: A study of sacred places. Journal of Environmental Psychology 24, 385-397, Elsevier
Osbon DK (1991) Reflections on the art of living: a Joseph Campbell companion. Harper Perennial, New York
Singh H (ed) (2007) The Sikh encyclopaedia [online]. Available at http://www.thesikhencyclopedia.com. Accessed 22 Mar 2015
Tabb P (2006) First principles: architecture of the unseen. The Academy of Sacred Architecture Studies, Denver, CO
Thomas B (1996) Spiritual path sacred place- myth, ritual and meaning in architecture. Shambhala, Boston
Timotio RA (2008) The sacred in architecture: a study of the presence & quality of place making patterns in sacred and secular buildings. Texas A & M University, College Station

Chapter 5
Neighborhood Planning: Approach in Improving Livability and Quality of the Life in the Cities

Gaurab Das Mahapatra

Abstract With the growing rate of urbanization in India from 19.9 % in 1971 to 30.02 % in 2001 and projected 38.21 % in 2021 (Vaidya C, Urban issues, reforms and way forward in India. Working paper No.4/2009-DEA, Department of Economic Affairs, Ministry of Finance, Government of India, New Delhi, 2009), there is an ardent need to adopt human value-based planning approaches in order to foster livability and quality of life in the neighborhoods of ever-expanding cities. Planning a neighborhood is not just arranging buildings for multiple functions. This needs deep insight of the everyday routine and the social and economic restraints of the society that divide people. The proper provision of amenities, living and working spaces, keeping in mind the above links, will lead to better-designed spaces which in turn lead to an interactive and a healthy society. Thus, the livability and quality of life in cities would reach a desired criterion. This paper deals with planning-oriented approaches, where options can be provided with respect to factors like social mix, affordability and adaptive use, level of interaction, spatial organizations, degree of enclosures, and sense of option. A city has to grow together – from economically weaker sections to higher income groups. Otherwise, only mere cities can be created and not "livable cities" (Evans P, Livable cities?: Urban struggles for livelihood and sustainability. University of California Press, Berkeley/Los Angeles, 2002). This paper focuses on a typical module which can be studied by developed, developing, and underdeveloped nations to create a basis for urban study to improve the quality of life.

[N.B.: The observation, analysis, and inferences discussed in this paper are a result of a series of case studies, questionnaire surveys, and situation analysis carried on by the author as a part of his undergraduate thesis during the year 2011–2012 and extended later as a basis of his research. These studies were conducted in Indian cities of Kolkata, Guwahati, Bhubaneswar, Old Delhi, Visakhapatnam, and Bhopal.]

G.D. Mahapatra (✉)
GITAM School of Architecture, GITAM University, Visakhapatnam, Andhra Pradesh, India
e-mail: gaurabdasmahapatra@gmail.com

F. Seta et al. (eds.), *Understanding Built Environment*, Springer Transactions in Civil and Environmental Engineering, DOI 10.1007/978-981-10-2138-1_5

Keywords Neighborhood • Society • Affordability • Livability • Adaptability • Planning

5.1 Introduction

If we identify actions rather than decisions as a principal focus of planning practice, the being effective in the world becomes a desired criterion. —John Friedmann

The concept of achieving an ideal neighborhood is analyzed by understanding the present issues or problems or challenges, defining a path or line of action to solve them, identifying the needs to be addressed in this process, and lastly giving the tentative result for this whole search. The whole study to increase livability has to invariably involve the four classes of society, i.e., HIG (higher income groups), MIG (middle income groups), LIG (lower income groups), and EWS (economically weaker sections). The government of India defines the four classes as per Table 5.1.

A series of questionnaire-based survey and aspect-based research on a sample of people from each of the four discussed classes was analyzed. This study is explained with reference to Tables 5.2, 5.3, and 5.4.

Table 5.1 Income slab for economic classes (Estimation of Housing Shortage 2007)

The classes	Monthly income slab (in rs)
HIG	0–3300
MIG	3301–7300
LIG	7301–14000
EWS	14000–above

Table 5.2 Basis of study for an ideal neighborhood

The mess	HIG	MIG	LIG	EWS
Dwelling unit	Luxurious	Comfortable	Satisfactory	Adjustable
Sense of owning in terms of property and space	Absolute	Passionate	Adjustable	Poor
Living conditions	Excellent	Hygienic	Satisfactory	Unhygienic
Sense of sharing	Partial	More	Satisfactory	Most
Personal domain	Maximum	Less	Optional	No
Economic viability	Maximum	Satisfactory	Less	Least
Hierarchy in cross subsidy	First	Second	Third	Last
Approachability by others	Least	Optional	More	Most

Table 5.3 The path to take for achieving an ideal neighborhood

The path	Design	Involvement	Understanding
With respect to issues regarding all classes	Role of an architect or planner or designer	Both public and private sector	Psychology and needs of the people

Table 5.4 The need of an ideal neighborhood and the path of search

The need	HIG	MIG	LIG	EWS
Living conditions	More amenities	Better means	Affordable luxury	Basic services
Sense of owning	To continue	To increase	To strengthen	To begin
Sense of sharing	To begin	To strengthen	To increase	To continue
The search	A better society altogether			

Fig. 5.1 Basis of study for improving livability and quality of life in cities (Sketch courtesy: Author)

Since they belong to an individual class, their psychology and socioeconomy are to be studied for improvement in quality of life. In Fig. 5.1, the first fashionable car depicts HIG, the cab depicts MIG, the scooter depicts LIG, and the rickshaw depicts EWS. Observing the obstacles, followed by a calculated attempt to solve it and suggesting an expected outcome, is the sole purpose of this paper.

5.2 Role of Social Mix in Improving Livability

According to an ideal design, if a person in the neighborhood wants to stay in a relative livable domain, he/she can stay well secluded. But if they want to interact, they should have ample opportunity to do so. Thus, it provides a sense of option. However, this can be judged only by a matter of time, because building is a timeless process (Alexander 1979). Livability refers to an urban system that contributes to the physical, social, and mental well-being and personal development of all its inhabitants (Timmer and Seymoar 2005). It is about delightful and desirable urban spaces that offer and reflect cultural and sacred enrichment (Timmer and Seymoar 2005). Thus, livability is suitability for human living (Merriam and Webster 2005).

This suitability is justified with the introduction of desired as well as required social mix in everyday life. Figure 5.2 illustrates the topic. The creeper grows around the tree and is supported by the tree. The tree has no problem in sustaining it; in fact, its beauty might increase due to its presence. The tree depicts the supply end of the urban society which is always encircled by parameters like social values, status, mixed approaches, housing shortage, public life, private interventions, economic barriers, and affordability (Masika et al. 1997). So, interdependency increases the level of mix and in turn helps to improve the livability in a society. Housing in particular, for all the four classes (HIG, MIG, LIG, and EWS), can be created in

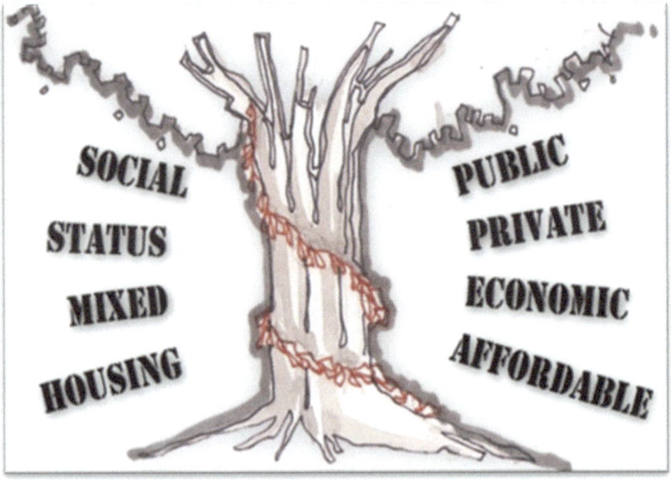

Fig. 5.2 Role of social mix (Sketch courtesy: Author)

such a way that neither feels segregated nor does the activities of one overlap on those of the others, unless necessary.

5.3 Relationship Between Affordability and Adaptive Use

Affordability and adaptive use are the two issues which are to be dealt primarily if we want to humanize the social life of our cities (Wheatley 2004). Affordability is a relative term as it varies from person to person primarily depending on his/her socioeconomic condition besides having other multiple contextual meanings (Mayank et al. 2012). The concept of adaptive use with relation to the affordability can be explained by Fig. 5.3a–c. All the rooms shown in these three figures are of the same size. Figure 5.3a depicts an HIG bedroom where only a bed and few bedroom accessories like wall tapestry, etc. are present. Figure 5.3b depicts an MIG bedroom where a computer table and dressing table are also squeezed in due to lack of space in their house. Figure 5.3c depicts a LIG bedroom where a few more elements have been added. So, in order to keep up with the social needs, human beings usually tend to modify their space for accommodating a larger sphere of life which in turn reduces the quality of life. Primarily, the needs are to be properly identified for addressing a proper design solution. This can be facilitated through case studies. For this paper, case studies and situation analysis were done to understand different strata of the society, and a comparative analysis is derived which can be referred in Table 5.5.

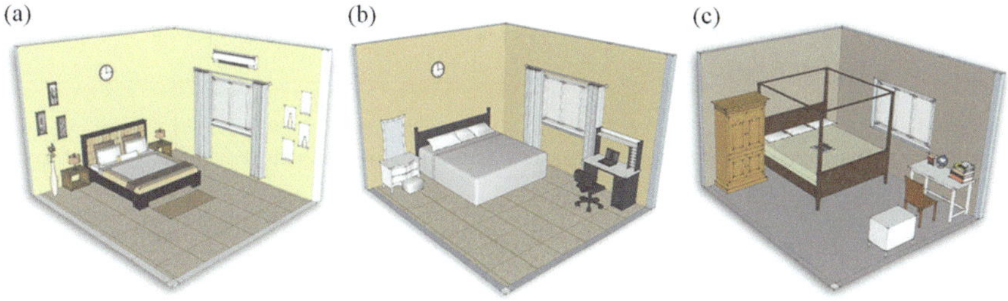

Fig. 5.3 Comparative spatial analysis (Sketch courtesy: Author). (a) HIG module, (b) MIG module, (c) LIG module

Table 5.5 Comparative analysis between different classes

Conditions	HIG	MIG	LIG	EWS
Nature of locality	Highly planned	Planned or semi-planned	Semi-planned or unplanned	Unplanned
Plot area	Huge	Adequate	Small	Confined to dwelling unit
Interaction with others	Nil or minimum	Moderate	Healthy	Maximum
Entrance	Defined	Satisfactory	Just for the sake of it	Common
Overlapping spaces	Minimum	Satisfactory	Adequate	Maximum
Space allocation	Specific uses only	Specific cum mixed use	Mixed use	Mixed use
Circulation	Through public zones only	Through public and some private zones	All zones	All zones
Aesthetics	Excellent	Good	Neat	Least
Use of racks	Showcase	Storage cum showcase	Storage	Multifunctional
Place for worship	Worship room	Small zone	Structural niches	Structural niches
Sanitation facilities	State of art	Hygienic	Semi-hygienic	Unhygienic
Amenities	State of art	Good	Acceptable	Minimum
Economic viability	Very strong	Strong	Good	Weak

5.4 Encouraging Level of Interaction Through Degree of Enclosures

> The essential thing of both room and square is the quality of enclosed space. —Camillo Sitte

Enclosure gives a sense of space, security, identity, and location. While designing an outdoor space, the character of the space gives an identity as well as helps in the mental mapping.

In Fig. 5.4 a–c, the different degrees of enclosures leading to different spatial organizations are shown, where each of them will have different impacts on the quality of life of an individual using them. The designer can improvise them by encouraging interaction through exploring different hierarchies of enclosures. Figure 5.5a–c will further explain this aspect. Articulated enclosures will promote – location and privacy, defining a territory, impart surveillance, decrease vandalism, and finally increase social contact (An introduction to Housing Layout, A GLC study 1980).

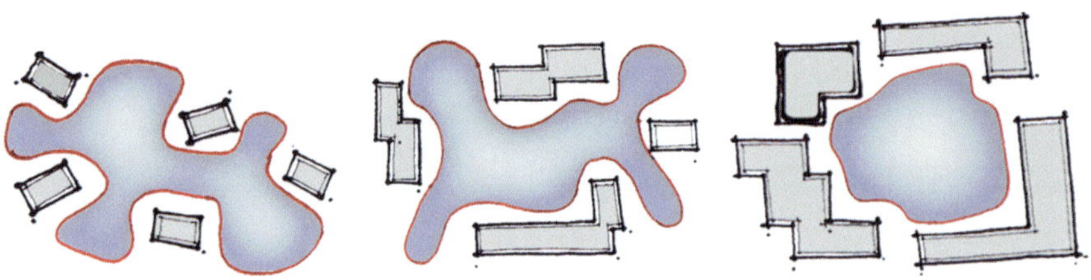

(a) (b) (c)

Fig. 5.4 Degrees of enclosures (An introduction to Housing Layout, A GLC study 1980). (**a**) Little, (**b**) Partial, (**c**) Strong

(a) (b) (c)

Fig. 5.5 Hierarchies of open space. (**a**) Intra tower, (**b**) Inter tower, (**c**) Inter block

5.5 Conclusion

It is often said, "*India is not a poor country, Indians are poor.*" As per statistics, top 5 % of Indians hold 20 % of India's economy (Das 2012). As a result, inequality of wealth is one of the age-old problems in India. Complemented by this are problems of population, lack of infrastructure planning, and immobilized government policies. There is a saying by Mahatma Gandhi that "*we need to change ourselves to change the world around us.*" But as an architect, the above sentence can be well framed by saying "the world can be changed by changing the architecture around it." Thus, my proposal is an attempt to understand the concept of urban habitat under the rich sociocultural context in developing nations. This attempt can only be fulfilled if we go on taking people as samples for research and analyze their spatial and psychological nature at regular intervals. As the society adapts itself to modernization with every passing year, the inferences will be liable to change with passing generations.

References

Alexander C (1979) Timeless way of buildings. Oxford University Press, New York

An introduction to Housing Layout, A GLC study (1980) Architectural Press, London

Das A The main factors of the lack of economical uniformity and balance in India. http://www.indiastudychannel.com/. 03 06 2012. http://www.indiastudychannel.com/resources/152882-How-can-we-bring-economical-uniformity-society.aspx

Estimation of housing shortage. Report of Technical Group, Planning Commission, Government of India, 2007, New Delhi

Evans P (2002) Livable cities?: Urban struggles for livelihood and sustainability. University of California Press, Berkeley/Los Angeles

Masika R, de Haan A, Bade S (1997) Urbanisation and urban poverty: a gender analysis. Swedish International Development Cooperation Agency (Sida), Brighton

Mayank H, Nanavaty M, Chakraborty SD, Lamaye A, Mitra S (2012) On point: affordable housing in India. Real Estate Intelligence Service, Jones Lang Lasalle

Merriam, and Webster (2005)

Timmer V, Nola Dr. Kate Seymoar (2005) The livable city. THE WORLD URBAN FORUM 2006 Vancouver Working Group Discussion Paper, Vancouver Working Group, Vancouver

Vaidya C (2009) Urban issues, reforms and way forward in India. Working paper No.4/2009-DEA, Department of Economic Affairs, ministry of Finance, Government of India, New Delhi

Wheatley CJ (2004) Humanization of social life, volume I- theory and challenges. Cultural heritage and contemporary change series VII, seminars on culture and values. Council for research in values and philosophy, Washington, DC, pp 112–115

Chapter 6
Interrelation of Public Open Spaces and Social Behavior: A Chronological Perspective

Sunaina Kapoor and Vanitha Putta

Abstract As Aristotle, the Greek philosopher and scientist, states, "man is by nature a social animal." And as evidence suggests, physical and social environment not only have an impact on human achievement but also fosters social interaction. A public open space not only authenticates the social behavior of a man irrespective of era, class, lifestyle, religion, political backgrounds, and economy but also is an integral part of human existence which inculcates the ethical and standard values. Since historical times, the provision of public spaces has been of great significance, and historical city planning clearly reflects importance given to the sense of community and provision of holistic interactive spaces to create a sense of security and belonging. The paper examines through a literature-based research study the significance of public open spaces in a chronological order from ancient times to the twentieth century. The paper gives an overview of interrelation between public open spaces and social interaction and the behavioral pattern of people with respect to the chronological study. The paper further identifies parameters, through the chronological study, to improve the public open spaces and social interaction in urban areas.

Keywords Public open spaces • Social behavior • Historical open spaces • Community relations • Recreational space • Interaction

6.1 Introduction

Public open spaces are diversely characterized as the open spaces which cater to all people in terms of space provision for recreation, development of social relations, interaction, or celebrating various festive occasions which are an integral part of Indian society. Public open spaces have been a boon for the nurturing humans in tangible and intangible manner. The development of a society depends on the vicious cycle of interdependency of human on public open spaces and vice versa.

S. Kapoor • V. Putta (✉)
Lovely School of Architecture and Design, Lovely Professional University, Phagwara, India
e-mail: kpr.sunaina@gmail.com; vanitha.ark@gmail.com

F. Seta et al. (eds.), *Understanding Built Environment*, Springer Transactions in Civil and Environmental Engineering, DOI 10.1007/978-981-10-2138-1_6

On one hand, public open spaces have an impact on human behavior pattern; furthermore, the people of society shape the pattern of public open spaces. Public open spaces need to be considered under limelight so as to understand their role in shaping of society. (The physiological values of open space)

6.2 Methodology

The methodology followed for the study is ethnography-based research. In order to study the interrelation of open spaces with human behavior pattern, a literature-based research is done to understand the open spaces during different time periods. Data collected is primarily through documented research and interpretation and analysis of the same with respect to various cultural and social groups of societies during the different time periods. The analyzed data is discussed in chronological order to understand how open spaces and behavior pattern have evolved with time. The ethnography study includes study of open spaces in cities of ancient time period till cities of the twentieth century. Further, a set of conclusions are derived by analyzing the need and impact of open spaces and social behavior on each other so that a utopia in urban scenario can be achieved and a need to determine the parameters derived from the past and incorporating them into the present and future.

6.3 Significance of Public Open Spaces

The evolution of humans as a part of community from the ancient times till now reflects the progression with respect to significance of community relations and transformation in the human behavior pattern which are influenced by various parameters. Such parameters include physical, socioeconomic, cultural, environmental, etc. Man, being a social animal, since historic times lived in groups primarily for a sense of security. Through times immemorial, this can be observed from the kind of structures man built for social gatherings or celebrations, and this in turn reflects the amount of importance given to the presence of public open spaces in the cities.

The formation of any public space in the communities/cities is governed by factors such as administrative system, religious beliefs, social and cultural aspects, and economical values of a community. Table 6.1 shows a brief comparison of historical examples and governing factors of ancient civilizations which echoes the significance given to public spaces and the influence of various factors.

(Fletcher 1999) (Great Buildings) (Ancient Civilizations History Website) (Athenian Agora Excavations) (History Learning Site) (Indus River Valley Civilization) (Schoenauer) (Landscapes, Greek) (Time Map) (Ancient Greece) (Landscapes, Greek) (Ancient Egypt Online)

Table 6.1 Comparison of ancient civilizations on the basis of identified parameters

	Governance	Social stratification	Status of women	Religious beliefs	Identified public open spaces
Ancient Mesopotamia 3100 BC–529 BC	○⊙	□	○	○	City streets and gate spaces (May 2013), landscaped gardens, palace and temple courtyards
Ancient Egyptian 600 BC–1900 BC	○	□	○	○	Temple plaza, monumental avenues
Ancient Indus civilization 2600 BC–1900 BC	◆	□	⊙	○	Great Bath, public wells, market places, streets
Ancient Greeks 510 BC–338 BC	⊙	□	⊙	○	Pnyx, agora, theaters, Olympia, stadiums
Ancient Romans 753 BC–476 BC	⊙	□	⊙	○	Forums, Colosseum, Circus Maximus, plaza, stadiums
Renaissance fourteenth to mid-seventeenth century	■	□	○	⊙	Public buildings like theaters around open spaces, landscaped gardens
Baroque 1600–1750	○	□	○	●	Landscaped gardens, parks, avenues, squares, plaza
Colonial	❖	□	○	⊙	Lack of open spaces in industrial zones. Mere presence of gardens or park
Industrial 1760–1840	❖	□	○	⊙	Lesser open spaces in industrial zones. Focus on development of shelters for masses

Note: Governance: ○ Monarchy, ■ Oligarchy, ◆ Egalitarian, ⊙Democracy, ❖ Multiple systems. Status of Women: ○Higher-class women had privileges, ⊙No right to vote, Social Stratification: □ Yes. Religious Belief: ○Polytheism, □Diverse religions, ●Christianity. Gallion and Eisner (2003)

Evidence suggests throughout the historical evolution of public open spaces there has been complex and shifting relationships between spaces and their functions and the social interaction. The public open spaces of current times and their form and purpose are certainly derived from the open spaces of ancient civilizations. Furthermore, these spaces have also evolved in terms of form and function with respect to access and user type and the kind of social interaction prevalent which, simultaneously, have been influenced by aspects of governance, social structure, gender, and religious beliefs which are intertwined.

Governance From monarchy to democracy, the governing systems' concern over power or divinity supported the development of public open spaces under their reign as a part of cultural, religious, or political needs. In Greek and Roman civilizations, the public spaces, predominantly characterized through their aesthetic features, were immensely valued, and they served the purpose of social interaction. The Greek *polis*, which emphasized the importance of public space to life, also played a crucial role in the politics and is often described as democratic space (aptly represented by *agora* in ancient Athens). The varied governmental systems, monarchs, or oligarchies in Greek civilization among different states influenced the open spaces and social interaction in terms of accessibility to people. The democratic public space had multiple functions, where citizens can interact and discuss issues pertaining to the city, such as used for commercial purposes and as an informal meeting place and community space. However, during the period of Renaissance which had the oligarchy governance, the public open spaces were centralized under the central rule of church located primarily in front of the church which also served as market place. Due to the location and function of the space, it thus became focal point for formal and informal public activities of the community. The civilizations with monarchy reflected public spaces which were designed to display power and wealth. During Baroque, practice of monarchy with growth of secularism in the governance shaped aesthetically and politically sound piazzas which were generally designed by artists and architects often hired by powerful authorities. In public open spaces, sense proportion and scale was understood and implemented to achieve grandeur public open spaces, which determined the governance authority. Contradictory to this was the Colonial period with multiple governing systems which included capitalism, socialism, utilitarianism, and communism. Although growth of market places brought commercial success, the content of health and grandeur of open spaces were minimal. The primary function of public space drifted from religious to commercial purpose. Specialization of the types of market places came into existence due to development of commercialization-oriented community, and these spaces formed into focal point of the urban community for interaction, buying and selling, and other formal and informal activities. Lack of centralized public open spaces and more development of industrial zones gave rise to slums and inappropriate living conditions which developed poor living conditions for the working class in general. The governance system was oriented for industrial development which did not incline with the protection and care of common people.

The mass migrants to cities in search of livelihood and shelter gave rise to quick development of less standardized housing, wherein the need for public interaction and provision of recreational spaces was deemed as less significant in comparison to demand for basic amenities and infrastructure. Due to increase in population, lack of basic infrastructure and public amenities developed a not so good living condition for the working class, while the upper classes enjoyed the luxury for better living (Carmona).

Social Stratification factor which has been responsible for the accessibility to open spaces which in turn governed by the governing systems typology and religious beliefs. During Greek civilization, with only one-seventh of population as citizens, non-citizens, i.e., the women, foreigners, and slaves, were restricted access to many public spaces. Such stratification and exclusion governed the belongingness of an entity to the unrestricted access to public open space.

Various public spaces were evolved to serve varied functions, *acropolises* for religious and secular assemblies and commerce and *agora* for daily communications and formal and informal assembly. As organic growth gradually gave way to planned urban form, the conscious design of public space increasingly reflected the notion that its aesthetic quality would impart an experience to the soul of its users (Carmona).

Distinct hierarchy of classes existed under every typology of governance system which suggests the influence of economy and social status that stratified the community in unequal categories (Kostof 1991).

Women of all classes were expected to run household and raise their children; privileges like freedom and rights for women to be active in social political attributes were present, while restrictions bounded the mind-set and lifestyle of people. Incorporation of laws for protection of women's right in practice suggests concern by society. Access to education was limited to higher-class women which was provided in restricted environments. Communities followed religion and had strong belief which was strongly visible in politics, culture, and living style of people. With the evolution of human civilization, space contestation came into existence due to increase in population and growth in requirements (Carmona).

6.4 Interrelation of Pubic Open Spaces and Social Behavior

6.4.1 Impact of Social Values on Public Open Spaces

The shaping of public open spaces is the resultant factor affected by political, religious, and cultural beliefs. The intangible attributes contribute to the shaping of public open spaces by development of spaces as the need of sharing or interaction; the divinity of the governance reflects in the finely designed public open spaces which cater the tangible and intangible needs of people. The belief and the interests of people and the mode of entertainment adopted by people suggest the framing of public open spaces. The sense of ethical values and moral responsibility raise a platform for the development of public open spaces.

6.4.2 *Impact of Public Open Spaces on Social Values*

Public open spaces found in urban scenario provide different social values on the basis of which people tend to interact and share thoughts. The sense of security and belongingness comes from thoughtfully scaled public open spaces. The rest in minds of community people and belief in the governing authority stay with the development of such spaces. The existing public open spaces inculcate a sense of freedom and responsibility embedded in community people which in turn develops the intellects of community as a prosperous entity and healthy community.

6.4.3 *Key Observations on the Influence of Factors on Transformation of Open Spaces*

Political Factor: The governance role has been very crucial since historical times. The key rules and regulations were evolved for physical space division and designation of use to the space division and designation of use to the space. The political factors transformed and retransformed open spaces and space in general which very often determined the distribution of power of politicians.

Social-Economic Factor: The class division which was apparent in each era in history. Although it evolved, the class division never vanished and reflected space segregation and access to open space based on income, caste, creed, race, and gender.

Cultural Factor: Same space is observed in various perspectives by different observers. Music, sports, dance, drama, and religious ceremonies invoked sense of belonging and acceptance in people.

Environmental Factor: The attention given to open spaces and efforts to preserve them has also progressed immensely with time.

6.5 Conclusion

The public open spaces have evolved with time and evolution of society. Historically, it is observed that space by nature is non-static and constantly changes with change in society and vice versa. Also it is observed that the meanings associated with these spaces change and evolve with time. Various factors such as political, social, cultural, economic, and environmental influence the public space transformation and uses associated with it which in turn influences the human behavior pattern.

People, bounded altogether by political hierarchy entangled with religious and cultural belief, engaged in survival of social, economic, and basic level from day-to-day life. Framing of laws and developing modes of punishment on not adhering to them reflects sense of responsibility and adherence to standard ethics expected to be followed by all so as to maintain the code of conduct and peace. Rebels and refinement of belief in people were witnessed in later times where freedom of expression was nurtured in every aspect. Specific concern toward economically weak sections was overlooked, but the survival persisted.

Development of art, literature, and technology flourished, and people of the communities experimented the never ending process that happened not on an individual level but in mass which existed due to the presence of interactive open spaces, sometimes planned and sometimes evolved.

Survival and nurturing of human community happens on the efficient and effective working of system as a single unit with complex yet organized interrelations. Public open spaces provide platform where people explore their potential by actively or passively involving in activities that contribute development of individual and in turn development of society. Human behavior varies with political background, religious belief, economic status, and social and cultural variables, while the magnitude of the behavioral pattern is directly interrelated and proportional to the efficiency of the public open spaces. Thus, historically, this interrelation is evident, and there is a need to address the concern toward analyzing and framing these parameters which are effected by or affect the public open spaces in order to create better communities, habitable spaces, and a better society.

References

Ancient Civilizations History Website (n.d.) Retrieved March 26, 2015, from Ancient Civilizations History Website: http://www.anciv.info

Ancient Egypt Online (n.d.) Retrieved March 23, 2015, from http://www.ancient-egypt-online.com

Ancient Greece (n.d.) Retrieved March 17, 2015, from http://www.ancientgreece.com/s/Life/

Athenian agora excavations (n.d.)Retrieved March 15, 2015, from http://www.agathe.gr/democracy/athenian_citizenship.html

Carmona M (n.d.). Retrieved September 10, 2015, from http://johannes.lecture.ub.ac.id/files/2012/09/Public-Space-_-The-Management-Dimension.pdf

Fletcher B (1999) A history of architecture. CBS Publishers and Distributors, New Delhi

Gallion AB, Eisner S (2003) The urban pattern – city planning and design. CBS Publishers and distributors, New Delhi

Great Buildings (n.d.) *Great buildings*. Retrieved May 28, 2015, from Great Buildings: http://www.greatbuildings.com

History learning site (n.d.) *How was Rome governed*. Retrieved March 24, 2015, from History learning site: http://www.historylearningsite.co.uk

Indus River Valley Civilization (n.d.) *Indus river valley civilization*. Retrieved March 26, 2015, from Indus River Valley Civilization: http://civilization.yolasite

Kostof S (1991) The city shaped. Thames and Hudson Ltd., London

Landscapes, Greek (n.d.) Greek landscapes. Retrieved March 15, 2015, from Greek Landscapes: http://greeklandscapes.com

May NN (2013) The fabric of cities: aspects of urbanism,urban topography and society in Mesopotamia,Greece and Rome. Brill Academic Publishers.

Recivilization (n.d.) *Recivilization*. Retrieved March 27, 2015, from Recivilization: http://recivilization.net

Schoenauer N 6,000 Years of housing

Staff H (n.d.) History. Retrieved March 17, 2015, from History: http://www.history.com
The Land Betwen The Rivers (n.d.) The land between the rivers. Retrieved March 27, 2015, from The Land Betwen The Rivers: http://landbetweentherivers.weebly.com
The physcological values of open space (n.d.) *Nora J. Rubinstein*. Retrieved March 27, 2015, from The benefits of open spaces: http://www.greatswamp.org/publications/rubinstein.htm
Tiesdell MC (2007) Urban design reader. Elsevier/Architectural press, Burlington
Time Map (n.d.) Time Map. Retrieved March 15, 2015, from Time Map: http://www.timemaps.com

Chapter 7
Provision of Ecosystem Services Through Urban Parks in Mumbai

Shruti Verma

Abstract Urban parks provide a range of ecosystem services to the urban population such as recreation, aesthetics, nature experience and education (cultural services) and microclimate regulation, air purification, rainwater absorption and noise reduction (regulating services). With a low open space per capita ratio, Mumbai City's high population greatly depends on urban parks to provide them with ecosystem services. However, these parks often face issues of poor maintenance, poor quality materials, restricted access and public safety. This paper explores the potential of urban parks in the city in providing ecosystem services to inhabitants.

Four neighbourhood parks in the western suburban region of Mumbai (H/West ward) have been studied to identify the ecosystem services provided by them. Data has been collected through interviews with visitors, observation and measurement to identify cultural and regulating services provided by the parks.

The paper concludes with a list of service-generating units/factors in urban parks that facilitates provision of ecosystem services at site level. This research would be useful for local authorities in design and management of parks in the city.

Keywords Ecosystem services • Urban parks • Service-generating unit/factor • Urban landscape

7.1 Introduction

Ecosystem services are the goods and services provided by the ecosystem to human beings. They range from provisioning (food, water, fuel, etc.), regulating (climate regulation, water purification, etc.), cultural (spiritual, recreation, etc.) and supporting services (primary production, soil formation, etc.) (MA 2005). In the urban context, green spaces are the main ecosystem provider at local level (Tratalos et al. 2007). With the evolving needs of growing populations in cities, open spaces are

S. Verma (✉)
IES College of Architecture, Mumbai, India
e-mail: shruti.verma@ies.edu

F. Seta et al. (eds.), *Understanding Built Environment*, Springer Transactions in Civil and Environmental Engineering, DOI 10.1007/978-981-10-2138-1_7

under immense development pressure. Urban inhabitants have to rely on the marginal open spaces in the form of natural features, playgrounds and parks that are available for interaction with nature.

Pertinent literature highlights the role of urban parks in microclimate regulation, air purification, noise reduction and enhancing air circulation around sites (Georgi and Zafiriadis 2006; Bernatzky 1983; Nilsson and Berglund 2006). Urban parks provide a platform for social interaction, give respite from city stressors such as crowds and noise pollution, aid personal development and thereby improve the quality of life. According to Bolund and Hunhammar (1999) and Breuste et al. (2013), as designed vegetated spaces, urban parks can provide the following ecosystem services:

- Cultural – aesthetics, cultural values, recreation, nature experience, learning about nature
- Regulating – air filtration, microclimate regulation, rainwater drainage, noise reduction
- Habitat – biodiversity

Mumbai is the most populous Indian city with a population density of 19,652 people per sq km (Census of India 2011). Mumbai has as little as 1.1 m^2 open space per capita which is 24 times lesser than New York and nearly 29 times lesser than London's open space ratio (Times of India 2012). In a study conducted by the MMR-EIS in Mumbai (Fig. 7.1), the open space distribution consisted of natural features (11 %), vacant sites (31 %), recreational open spaces (35 %) and open spaces with built structures (23 %). These statistics highlight the pressures of urbanisation on open spaces as a considerable portion (23 %) are encroached upon with built structures and only 11 % are available for natural features.

Recreational open spaces have the largest share of 35 % in the open space distribution of Greater Mumbai. These reserved gardens, parks, playgrounds and recreation grounds are maintained by the Garden Department of the Municipal Corporation of Greater Mumbai (MCGM). Though efforts are taken to provide facilities to residents through the planning and design of these spaces, they face

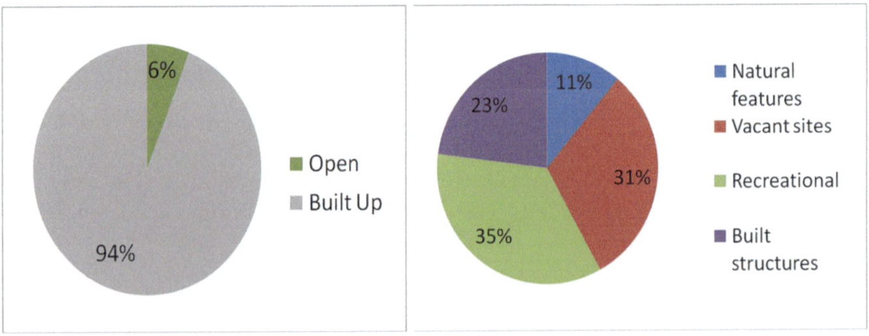

Fig. 7.1 (*Left*) Land reserved as open space and (*right*) actual open space distribution in Greater Mumbai (Based on data of MMR-EIS 2012)

issues of visibility, public safety, poor maintenance and incongruous usage of parking and garbage dumping. Further, many of these recreational spaces are not in use as they are either occupied by built structures, not developed for use or have restricted access (MMR-EIS 2012).

7.2 Study Area

Study area chosen is H/West ward which comprises localities of Bandra, Khar and Santacruz (West) in the western suburbs of Mumbai. Total area of ward is 11.55 km^2 with a population density of 36,500 persons/km^2. The ward has 85 reserved open spaces – 14 gardens, 41 recreation grounds and 30 playgrounds (Open (Open and "Open Mumbai: Re-envisioning the City and Its Open Spaces" 2010)). Though 90 % of the ward is within 250 m distance of an open space (MMR-EIS 2012), many of these open spaces have restricted access as they are institutional greens or gymkhanas/clubs, thereby reducing ecosystem services provided to the residents (Fig. 7.2).

Figure 7.3 shows the locations of the parks selected for study: Almeida Park (P1), Arunkumar Vaidya Udyan (P2), Jogger's Park (P3) and Rotary Park (P4). Figure 7.4 shows views of the parks selected for study.

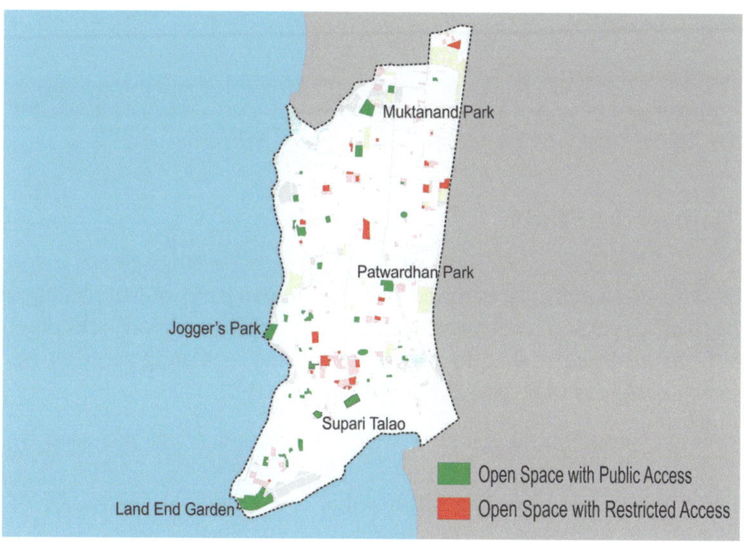

Fig. 7.2 Map of H/West ward showing open spaces with restricted and unrestricted access (Source: Author)

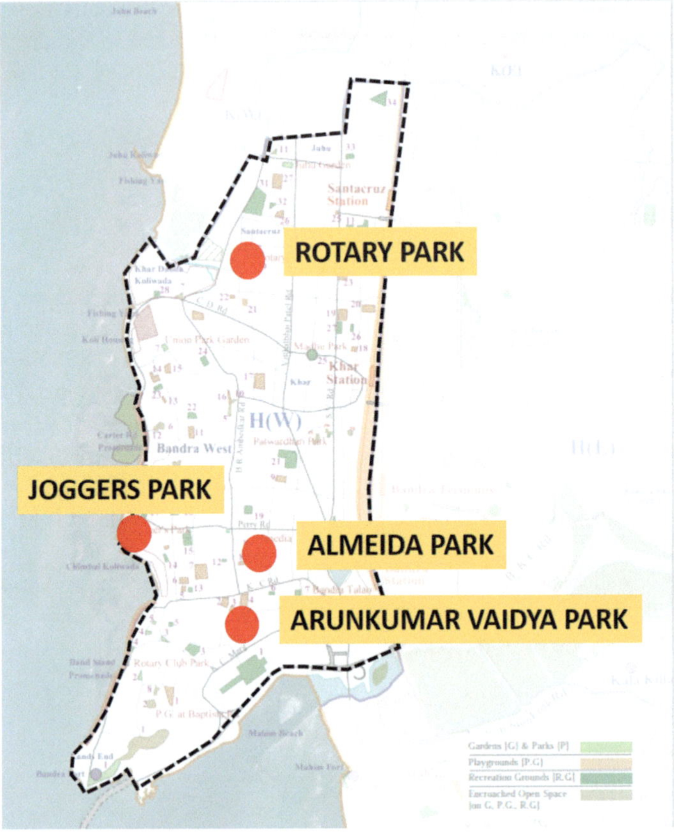

Fig. 7.3 Map of H/West ward of Mumbai showing location of park case studies (Base map: Open (Open and "Open Mumbai: Re-envisioning the City and Its Open Spaces" 2010))

7.3 Aim of the Study

The aim of this research is to examine the role of urban parks of Mumbai in providing ecosystem services at local level to inhabitants. Neighbourhood parks of medium size (5–15 ha area) are studied to assess cultural and regulating services. Some of the research questions to be answered are:

- In what way are cultural and regulating ecosystem services provided by these parks?
- What are the factors or design features that contribute to the provision of these services?
- Does spatial distribution of such parks in the given area give rise to equitable distribution of ecosystem services?

Fig. 7.4 Photographs of parks under study. (**a**) Almeida Park, (**b, c**) Arunkumar Vaidya Udyan (MET)-maintained portion and MCGM-maintained portion, (**d**, **e**) Jogger's Park and (**f**) Rotary Park (Photographs source: Author)

7.4 Methodology

The methodology was based on usage characteristics and activity patterns to study cultural services and field measurements, the presence of indicators and user responses for the regulating services. Multiple daily and weekly site visits were made during over a 2-month period, and data in the form of visual material (activity pattern maps, photographs), environmental variables (air temperature, wind speed, sound levels) and interview-based surveys was collected.

The survey addressed demographic information, use of park (such as frequency of visits, distance travelled) and visitor preference. Two hundred people from different age groups and social classes were randomly selected and interviewed for the research.

Observation method was used to identify the presence of indicators for regulating services listed in Table 7.1. Air temperature and wind speed were measured on site using an anemometer. Recordings were taken at critical points such as along the park edge, surrounding roads, centre of the park and in surrounding areas, at a distance of around 50–100 m from the park boundary at 1.5 m above ground level. Sound levels were measured using Sound Meter v2.8 at the entry of park, along the park edge, surrounding roads and in the centre of the park.

Table 7.1 List of indicators used for identification of regulating ecosystem services

S. no.	Indicator	References
1.	*Air purification*	
(i)	Dust accumulation on leaves	Crow (2011), Hiemstra et al. (2008)
(ii)	Large trees on edge of park	Center for Urban Forest Research (2004), Nowak (1994)
(iii)	Increased tree cover	Nowak et al. (2006)
(iv)	Visitors' experience of better air quality to surrounding areas	O'Leary (1988)
2.	*Microclimate regulation*	
(i)	Climatic conditions of park being a reason for visit	Lin et al. (2013)
(ii)	Reduction in temperature between surroundings, edge of park and centre of park	National Recreation and Park Association (2010)
(iii)	Increased park area under tree canopy cover	Souch and Souch (1993)
(iv)	Increased unpaved areas in the park	Henry and Dicks (1987)
(v)	Nocturnal airflow from park	Eliasson and Upmanis (2000)
3.	*Rainwater absorption*	
(i)	Reduced impervious ground cover in the park	Haase and Nuissl (2010)
(ii)	Storm water drains of surrounding areas open into park	
4.	*Noise reduction*	
(i)	Reduction in sound levels between surroundings, edge of park and centre of park	Chronopoulos et al. (2002)
(ii)	Closely spaced vegetation at the park edge	Onder and Kocbeker (2012)
(iii)	Reduced visibility of sound source from inside the park	Aylor and Marks (1976)
(iv)	Multi-storey vegetation	Onder and Kocbeker (2012)

7.5 Analysis

Demographic information and user information such as age groups, distances travelled and durations of visit were studied in conjunction with other user data such as purpose of visits. Observations regarding the physical characteristics of the park and activities were used to validate the results. Environmental variables like changes in temperatures, wind speed and sound levels were analysed with respect to the physical characteristics such as percentage of permeable ground cover, ground cover materials, vegetation pattern and external factors such as traffic, construction activity and surrounding built form.

A comparative analysis between the four parks based on user response (in the case of cultural service) and the presence of indicators (in the case of regulating service) was performed. The service-generating unit or factor was concluded for each ecosystem service (Table 7.2).

7.6 Results

The interview results showed that the parks were used fairly equally by both genders. In parks where majority of visitors came from areas closer to the park, i.e. less than 500 m (P2, P4), there was an increase in frequency of usage and significant use of walking as mode of commute. Visitors travelled longer distances by vehicle to access parks (P1, P3) due to the absence of similar parks in their neighbourhood, which led to lesser frequency of usage. Parks that had playing facilities (P2) were found to have majority of visitors that belonged to the under 25 age category and were used for longer duration (above 2 h). Parks that were used primarily for walking/jogging (P3) were found to be used for shorter duration (up to 1 h).

7.6.1 Cultural Services

7.6.1.1 Recreation

Recreation in the form of physical activity or rest and relaxation was found to be the most common purpose for visit to the parks (see Fig. 7.5). Parks with well-maintained tracks of adequate length (P3, P4) were found to be widely used for walking/jogging (72 % and 64 %). All parks had play equipment for children and were used mainly in the evenings or after school hours by children. The parks were not used as much for exercise (16–32 %) due to the absence of unobstructed and designated space for the same. The presence of good tree cover, adequate provision of seating, sea face setting and proximity to educational institutes led to parks being used for rest and relaxation (P1, P3).

Table 7.2 Service-generating units and factors associated to ecosystem services provided by the parks

Ecosystem service		Service-generating units [U]/factors [F]
(A)	*Cultural services*	
1.	Recreation	**[U]** Provision of facilities for recreation such as vegetation and safe space for physical activity
		[F] Parks provide access to public for free or at a nominal charge for daily recreation needs
		[F] Parks located close to residential or educational areas (less than 500 m distance)
i.	Physical activity	
	Walk/jog	**[U]** Walking and jogging tracks
		(a) Minimum length of 350–400 m
		(b) Multiple tracks to avoid congestion and increase length
		(c) Suitable material for physical health
		(d) Inclination
		(e) Shading of track
		[F] Presence of residential settlements near the park
		[F] Safe environment for walking/jogging – proper infrastructure and security
	Play	**[U]** Play areas
		(a) Designated play areas for children of different age groups to avoid conflict
		(b) Large unobstructed space for sport activities
		[F] Parks located close to residential or educational areas
		[F] Safe environment for playing – proper infrastructure and security
	Exercise	**[U]** Space for exercise
		(a) Unobstructed soft surface for exercise such as a lawn and ground
		(b) Adult exercise equipment
ii.	Rest and relaxation	**[U]** Vegetation – provision of shade and relaxing environment by vegetation **[U]** Physical infrastructure – adequate number and maintenance of seating
		[F] Public parks provide visitors with opportunities for retreat from the dense built urban environment
		[F] Safe environment
		[F] Park timings – parks open till late morning are frequented by young college students for rest and relaxation
2.	Aesthetics	**[U]** Vegetation
		[U] Maintenance of park
3.	Nature awareness and education	**[U]** Presence of fauna in the park for young children
		[F] Growing disconnect of urban dwellers with nature in the city

(continued)

Table 7.2 (continued)

Ecosystem service		Service-generating units [U]/factors [F]
4.	Cultural values	
(i)	Meeting friends/ relatives	**[U]** Provision of space for activities such as sports training that enable group play and communication
		[F] Presence of residential settlements near park
		[F] Safe environment
(ii)	Social gatherings	**[F]** Park rules such as prohibition of social events and activities such as laughter clubs ensure a peaceful environment for visitors but also are a deterrent to social bonding
		[F] Presence of residential settlements near park
		[F] Safe environment for gatherings
(B)	*Regulating services*	
1.	Air purification	**[U]** Presence of large trees with trunk of large diameter and broad leaves for maximising air pollutant absorption
2.	Microclimate regulation	**[U]** Vegetation – shading, reduction in air temperature by evapotranspiration, reduction in surface temperature due to lower albedo level of vegetated ground cover
		[F] Urban heat island effect in city – heat due to high built-up area and anthropogenic heat
3.	Rainwater absorption	**[U]** Unpaved surfaces increase storm water absorption, reducing surface run-off and causing ground water recharge
4.	Noise reduction	**[U]** Presence of multi-storey and closely spaced vegetation at edge of park reduces noise levels
		[U] Partial visual barrier between sources of noise
		[U] Vegetated buffer between edge of park and user areas
		[F] Presence of high vehicular traffic route next to park increases surrounding area's sound levels

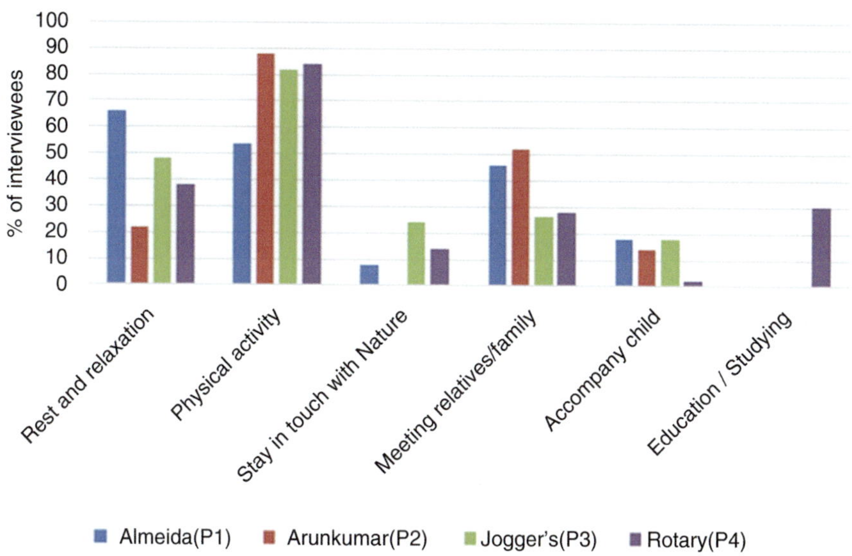

Fig. 7.5 Purpose of visit to the parks

7.6.1.2 Aesthetics

Parks in which visitors expressed satisfaction with the greenery were considered to contribute to the aesthetics of the area (see Fig. 7.6).

7.6.1.3 Nature Awareness and Education

Nature awareness and education was found to be absent in most parks under study (P1, P2, P4). However, through observation of fauna in P3, visitors believed there was a significant contribution to their awareness about nature (46 %).

7.6.1.4 Cultural Values

Parks that were used primarily for rest and relaxation (P1) or had provision for sport activities (P2) were found to be used more for socialising. Social gatherings such as laughter clubs, yoga group sessions, group gatherings of senior citizens and slum women were observed in the parks. Park rules were found to be a deterrent in social gatherings. No religious activities took place in any of the parks. Figure 7.7 shows cultural services provided by the parks under study.

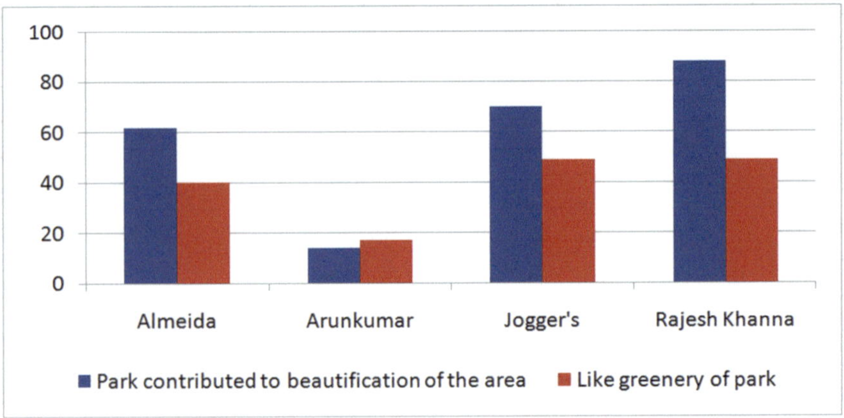

Fig. 7.6 Visitors' opinion on contribution of park to beautification of area and satisfaction with greenery of park

7.6.2 Regulating Services

7.6.2.1 Air Purification

P1 and P4 showed strong presence of air filtering ecosystem service due to the presence of all indicators (see Table 7.1, Fig. 7.8). Visitors' experience of air quality in P3 was found to be good (86 %) due to the effect of the sea.

7.6.2.2 Microclimate Regulation

Temperature difference between centre of park and surrounding areas was found to be between 0.8 and 1.4 °C in mornings and 0.5–1.6 °C in the evenings across all parks. Despite the absence of sufficient tree cover, P3 was found to be cooler than

Fig. 7.7 Cultural services provided by parks – duck pool in Jogger's Park (*left*) and schoolchildren playing under a tree in Almeida Park (*right*)

Fig. 7.8 Regulating services provided by parks – multi-storey vegetation along park edge for effective noise reduction (*left*) and dust accumulation on leaves (*right*) in Rotary Park

the surrounding area due to prevalence of strong breeze (0.5–1.5 m/s). There was no evidence of nocturnal breeze being generated from any of the parks, and this was validated with residents living around the parks.

7.6.2.3 Rainwater Absorption

All parks had ground surfaces that were minimally paved (less than 30 % of park area) which results in higher infiltration levels. P3 also provided rainwater drainage to the surrounding residential area as opening of storm water drains into the park was observed.

7.6.2.4 Noise Reduction

Reduction in noise levels in the range of 5–20 dB between surrounding road and centre of the park was found in all cases. P4 was found to have a strong presence of noise reduction ecosystem service due to the presence of all indicators (see Table 7.1, Fig. 7.8).

Based on the above results, Table 7.2 lists the ecosystem service-generating units/factors useful for design of urban parks.

7.7 Conclusion

7.7.1 Enhancement of Ecosystem Service Provision

Urban parks in Mumbai are found to be providers of various ecosystem services to city inhabitants but to a limited extent. These services can be enhanced by addressing the associated ecosystem generating units. For example, by providing space and environment for fauna, display boards with environmental awareness messages and organisation of environmental education activities with help of local resident associations and schools, nature awareness and education provided would be enhanced. Also, as parks are used on a regular basis, they should meet the daily needs of the population. Visitors expect good maintenance, security, greenery and basic water and toilet facilities from the park.

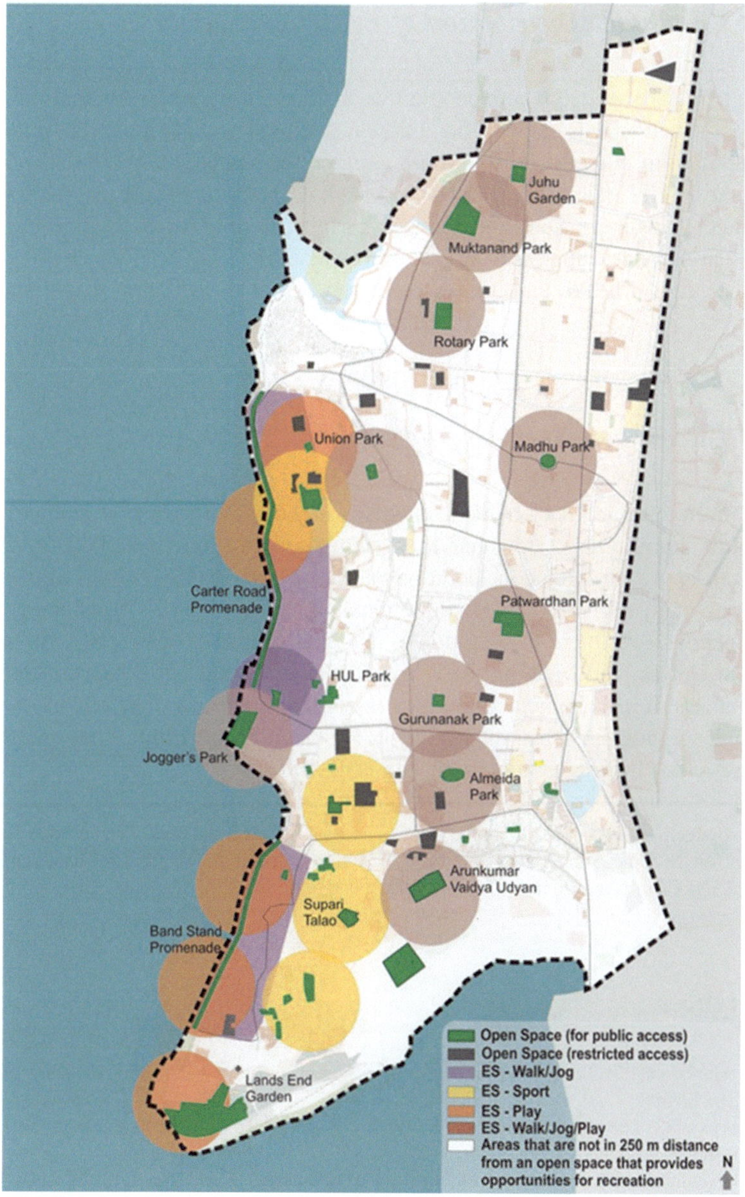

Fig. 7.9 Cultural ecosystem services provided by open spaces in H/West ward of Mumbai

7.7.2 *Equitable Distribution of Ecosystem Services*

Urban parks are needed where people live and need to be accessible to all in order to meet the daily needs of recreation of the residents. There is a good distribution of open spaces for playing and walking/jogging in the southern parts of the ward but not in the central areas of the ward (Fig. 7.9). It is important to consider the services provided by an open space and their access while planning for them. By integrating ecosystem services approach at local-level planning and decision making, key issues of human health and quality of life and climate regulation would stand to benefit.

7.7.3 *Ecosystem Services Approach*

The conventional approach to decision/policy making has focused on a single sector and often compromised on other ecosystem services (Tallis and Polasky 2009). Thus, an approach with consideration of ecosystem services would allow local authorities and decision makers to anticipate consequences of decisions/policies and potential investment in alteration of a natural feature or open space. As seen in cases in Durban and Melbourne (TEEB 2011), assessment of ecosystem services and analysis of management and policy options with various stakeholders can lead to viable solutions in urban management. Research in valuation of ecosystem services provided by urban parks in the city is needed.

Acknowledgement I would like to thank Prof. Meenal Sutharia and Prof. Vinit Mirkar for their guidance on the topic. Sincere thanks to Prof. Pratheek Sudhakaran, Dr. Shivika Chandra and Ms. Tanvi Manglik for their help in modelling, editing and data collection.

References

Aylor, Marks (1976) Perception of noise transmitted through barriers. J Acoust Soc Am 59(2):397–400

Bernatzky (1983) The effects of trees on the urban climate. In: Trees in the 21st century. Academic Publishers, Berkhamster, pp. 59–76

Bolund, Hunhammar (1999) Ecosystem services in urban areas. Ecol Econ 29(2):293–301

Breuste J, Schnellinger J, Qureshi S, Faggi A (2013) Urban ecosystem services on the local level: urban green spaces as providers. Ekologia 32(3):290–304

Census of India (2011) Mumbai City district: census 2011 data. [online] Available at: http://www.census2011.co.in/census/district/357-mumbai-city.html. Accessed 27 Feb 2015

Center for Urban Forest Research (2004) The large tree argument. [online] Available at http://www.fs.fed.us/psw/programs/uesd/uep/products/cufr_511_large_tree_argument.pdf. Accessed on 1 Mar 2015

Chronopoulos JA, Tsiotsios KM, Balotis G, Papafotiou M (2002) The impact of design on traffic noise control in an urban park. In: International Conference on Urban Horticulture 643, pp. 277–279

Eliasson I, Upmanis H (2000) Nocturnal airflow from urban parks-implications for city ventilation. Theor Appl Climatol 66(1–2):95–107

Georgi NJ, Zafiriadis K (2006) The impact of park trees on microclimate in urban areas. Urban Ecosyst 9(3):195–209

Haase D, Nuissl H (2010) The urban-to-rural gradient of land use change and impervious cover: a long-term trajectory for the city of Leipzig. J Land Use Sci 5(2):123–141

Henry JA, Dicks SE (1987) Association of urban temperatures with land use and surface materials. Landsc Urban Plan 14:21–29

Hiemstra JA, Schoenmaker-Van der Bijl E, Tonneijck AEG, Hoffman MHA (2008) Bomen: Een verademing voor de stad. Plant Publicity Holland [etc.]

CROW (Centrum voorRegelgeving en Onderzoek in de Grond-, Water- en Wegenbouw en de Verkeerstechniek), (2011) Solve - luchtkwaliteit en verkeer

Lin, Chuang-Hung, Tzu-Ping Lin, and Ruey-Lung Hwang (2013) Thermal comfort for urban parks in subtropics: understanding visitor's perceptions, behavior and attendance. Adv Meteorol 2013

Millenium Ecosystem Assessment (2005) Ecosystems and human well-being, vol 5. Island Press, Washington, DC

MR-EIS (2012) Inventorisation of open spaces and water bodies in Greater Mumbai for MMR-EIS

Nilsson ME, Berglund B (2006) Soundscape quality in suburban green areas and city parks. Acta Acustica United Acustica 92(6):903–911

Nowak DJ (1994) Air pollution removal by Chicago's urban forest. Chicago's urban forest ecosystem: results of the Chicago urban forest climate project, pp. 63–81

Nowak DJ, Crane DE, Stevens JC (2006) Air pollution removal by urban trees and shrubs in the United States. Urban For Urban Green 4(3):115–123

O'Leary D (1988) Air quality in the National Parks: a summary of findings from the National Park Service Air Quality Research and Monitoring Program, US Dept. of the Interior, National Park Service, Air Quality Division

Onder S, Kocbeker Z (2012) Importance of the green belts to reduce noise pollution and determination of roadside noise reduction effectiveness of bushes in Konya, Turkey. World Acad Sci Eng Technol 66:639–642

Open Mumbai (2010) Open Mumbai: re-envisioning the City and its Open Spaces

National Recreation and Park Association (2010) Air quality effects of urban trees and parks. [online] Available at http://www.nrpa.org/uploadedFiles/nrpa.org/Publications_and_Research/Research/Papers/Nowak-Heisler-Summary.pdf. Accessed on 4 Mar 2015

Souch CA, Souch C (1993) The effect of trees on summer time below canopy urban climates. J Arboricult 19:303–303

Tallis H, Polasky S (2009) Mapping and valuing ecosystem services as an approach for conservation and natural resource management. Ann N Y Acad Sci 1162(1):265–283

TEEB (2011) TEEB manual for cities: ecosystem services in urban management

Times of India You have just 1.1 square metres of Open Space. [online] Available at http://timesofindia.indiatimes.com/city/mumbai/You-have-just-1-1-square-metres-of-open- space/articleshow/13585198.cms. Accessed: 27th Feb 2015

Tratalos J, Fuller RA, Warren PH, Davies RG, Gaston KJ (2007) Urban form, biodiversity potential and ecosystem services. Landsc Urban Plan 83(4):308–317

Chapter 8
Patterns of Flow: The Spatial Dimension of Water in the Desert

Meghal Arya

Abstract The richness of the architecture for water in the desert emerges from a shortage of water that traditionally gave rise to ingenuous techniques to capture and store rainwater, and replenish, and source ground water. Each drop was precious. People used water in their daily chores judiciously. This paper will demonstrate that the spatial expression of water was closely interwoven in the lives of people through the making and practice. The artifacts are not isolated entities, but are rooted within living societies, shaped by the natural environment and active participants of spatial urban structures, contributing to an ecosystem that sustains life in the harsh climate.

The study emerges from a concern about how we "view and perceive" our resources. Modern infrastructure for water, being hidden from view and dissociated from experience, has erased from our urban sensibilities the connection with water and the environment. The components that bring water have been converted into monofunctional infrastructure. They perform a singular task unlike the historic water bodies that were multifunctional. The erasing from public memory, the spatiality of water, has taken away fundamental relationships of society, environment, and technology. To balance the homogenizing and monocultural trends of the contemporary world, this paper will contribute to make available the knowledge of the historic water system to initiate a productive dialogue toward an urban regeneration that is ecological.

Keywords Architecture for water • Multifunctional infrastructure • Urban regeneration • Conservation • Ecology

M. Arya (✉)
Faculty of Architecture, CEPT University, Ahmedabad, India
e-mail: meghalj@yahoo.com

F. Seta et al. (eds.), *Understanding Built Environment*, Springer Transactions in Civil and Environmental Engineering, DOI 10.1007/978-981-10-2138-1_8

8.1 Introduction

The architecture for water in the desert regions is remarkable in its diversity and richness. This architecture carves out spaces that integrate the knowledge and cumulative wisdom of centuries. Over time, we have tended to "see" these structures as isolated, independent edifices of quality craftsmanship. However, as water traverses the natural and human terrains, it creates strong physical relationships. The hydraulic structures became locations of sociocultural activity, environmental harmony, and tectonic innovations and expressed these relationships of human ecology (Fig. 8.1). As the sine qua non of any human settlement, water structures generated the first public "infrastructure" of any settlement. This paper discusses how the public multifunctional "infrastructure" moved from being linked, connected, and integrated to being monofunctional, completely disconnected in terms of tactile experience, spatial presence, and design integration. Illich maps this shift in our perceptions of water as a purifying element to a matter that causes contamination and itself needs purification (1985). This 180° turnaround in the manner of association with water has had implications on its spatiality and infrastructure. This paper, at a theoretical level, alludes to Illich's "question concerning the recuperation of dream water by the city child" (1985). The dissociation with water as an element accentuates major concerns of urban regeneration particularly related to "comprehensive and integrated vision and action" (Roberts and Skyes 2000) in the context of environmental and social integration. The first step is to comprehend the

Fig. 8.1 The Chand Baori at Abhaneri is a subterranean hydraulic structure to access subsoil water

infrastructure for water as an ecosystem. A study of the water infrastructure of Jodhpur provides an example of the concept.

8.2 Jodhpur Water System

In the arid regions of India, despite being water deficit, beautiful settlements from Jaisalmer to Shekhawati sprung up. They invested deeply in water harvesting and a diversity of types, dependent on regional geography emerged. Anil Agarwal and Sunita Narain (1997) and Anupam Mishra (2001) have mapped, researched, and documented many of these structures, though most of the work has been in the rural hinterland. Empirical evidence indicates that the water bodies were also closely entwined in the city fabric as seen in Jodhpur. Jodhpur, located on the border of the Thar Desert where it meets the Aravallis, is a thriving trading city, second largest in Rajasthan, and one of the most significant settlements in the desert.

The water system of Jodhpur consisted of several elements for collecting and storing water. These were large, medium, and small reservoirs that were constructed in the undulating topography of the Aravalli Range with a formation that constitutes valleys and ridges that are fragmented. A continuum between the various topographies is created through these reservoirs linked to the urban fabric. Nature and human-made are interrelated into an integrated experience through the architecture of ghats, embankments, gates, canals, and spillways. The steps on the embankment stabilize the wall of the dam, provide access to water as levels fluctuate due to irregular rainfall, express the craftsmanship of chiseled stone, and create an urban public space. They allow intimate small gathering on the landings, the frequency of which accounts for the load of several pots of water carried by the women on their heads. Technological know-how and keen observation are combined with sociocultural requirements of the settlement.

Apart from the 70 odd reservoirs, there also exist tanks, jhalaras, and more than 250 wells. With more than 350 water structures of varying types and scales, Jodhpur presents a complex water system. "Jodhpur is perhaps the only city in India where serious efforts had been made to conserve every drop of rain water ever since the city was founded" (Mohnot NA). Jodhpur sits on a rock formation that is "friable and soft nature [leading] to formation of small cavities in saturated zones. This makes it a very good aquifer forming chief source of ground water in the area" (Ground Water Board 2008). Water collected in the reservoirs moves through the cavities, seeping into the wells, jhalaras, and tanks. A hierarchy of architectural types used in conjunction with geological conditions creates the transport network of the water infrastructure that was diverse, appropriate to scale and location, and in harmony with nature with minimal effort against the land. The location of the water structures within communities, directly meshed in the daily lives of the people, reduced effort to travel long distances. This was possible by a thorough study of the geography. The resultant was a balance between social and community space, personalized access, and environment.

The places for water became the social heart of the city with easy and direct access. Even the smallest wells had platforms around them to facilitate social interaction. The infrastructure was created keeping in mind the large number of people likely to access the facility, considering the presence of elaborate steps and platforms leading to the water. These were gestures of sensitivity and pragmatic considerations of urban social conditions. The pyramidal structure of society implies more numbers of less privileged people at the bottom. The economically weakest sections performed many domestic chores in public areas including bathing. The most public water spaces are therefore the largest, with maximum potential of access – wider steps, more landings, and platforms. It is a mature society that acknowledges the social segregation and works to reduce it and rationalize it through mechanisms that bring the best to all. Some aspects of social structure are assimilated in the infrastructure. The 2011 Census suggests about 3,000 households still access water from the various tanks in the city.

Jodhpur demonstrates how the infrastructure network for water was closely interwoven in the lives of people through space and practice, making it multifunctional, diverse, and people-centric (Fig. 8.2). The elements of the system are not independent and isolated but rooted within living societies, affected by the natural environment, and are active participants of urban spatial organization, contributing to an ecosystem that sustains life in extreme climates.

8.3 Learning from History as a Strategy for Urban Regeneration

Current strategies for urban regeneration range from reinvigorating historic districts, leveraging unique cultural heritage, encouraging investment and development toward housing and cultural hubs, creating employment opportunities, etc. Economic implications are prioritized. In Indian cities, the past decade has seen a focus on the introduction or upgradation of infrastructure. Design determinants are primarily numerical and statistical efficiencies. This has taken infrastructure on a development curve that distances it from social conditions and inclusive design that considers cultural, economical, environmental, and technological factors. There is a concern about how we "view and perceive" our resources. Piped and cloaked under layers of earth, the movement of water and its collection and distribution are eliminated from our urban sensibilities. The components that bring water have been converted into monofunctional infrastructure. The system performs a singular task – to store and transport water unlike the historic water system that was multifaceted, integral, and inclusive. This paper opens up the discussion on the inclusion of sociocultural and geographical aspects in infrastructure development as a strategy for urban regeneration.

The water structures were components of a web of relations, creating a spatial ecosystem that facilitated living in the desert. The system integrates the diverse aspects of a society through relationships of mutuality, interdependencies, and networks

Fig. 8.2 The Ranisar at Jodhpur collects water from the surrounding hills by damming a narrow part of the valley. The bund, with its steps to access water, becomes a public space for the city. It provides water to the fort seen in the background through a system of wells, pulleys, and channels

(Fig. 8.3). The relationship with the environment is shaped by the empirical knowledge gained over time, passed from generation to generation along with the social and cultural aspects. These geo-cultural relationships traditionally developed assuming a synthesis and connection of systems.

8.4 Conclusion

The erasing from public memory, the spatiality of water, has eroded fundamental relationships between society, environment, the making, and other aspects of human endeavor, and the historic system languishes (Figs. 8.4 and 8.5). Though recent

Fig. 8.3 Soorsagar at Jodhpur; a careful assessment of the land combined with knowledge of hydrology where several types of water bodies were interlinked to create a system of taking water to the settlement – a process that persisted for more than a thousand years, showing variations in the manifestations, but continuity in the principle

Fig. 8.4 Mahilabaug Jhalara at Jodhpur; like all the other water structures, this beautiful structure lies abandoned and neglected though it continues to provide water to the city. Lack of care has made it a place for garbage leading to a loss of water quality

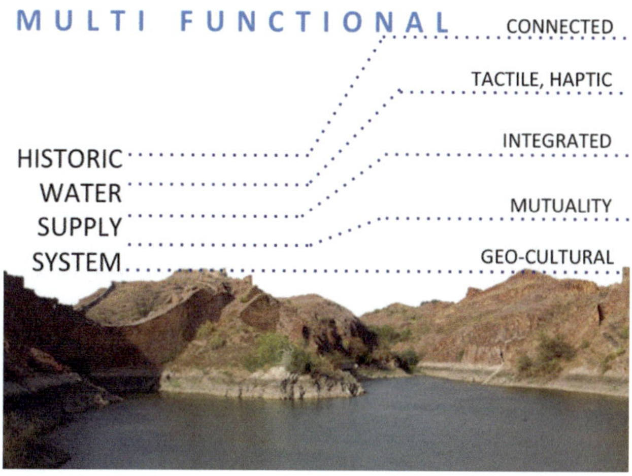

Fig. 8.5 The historic water structures were part of an infrastructure system that was multifunctional making it people-centric and humane

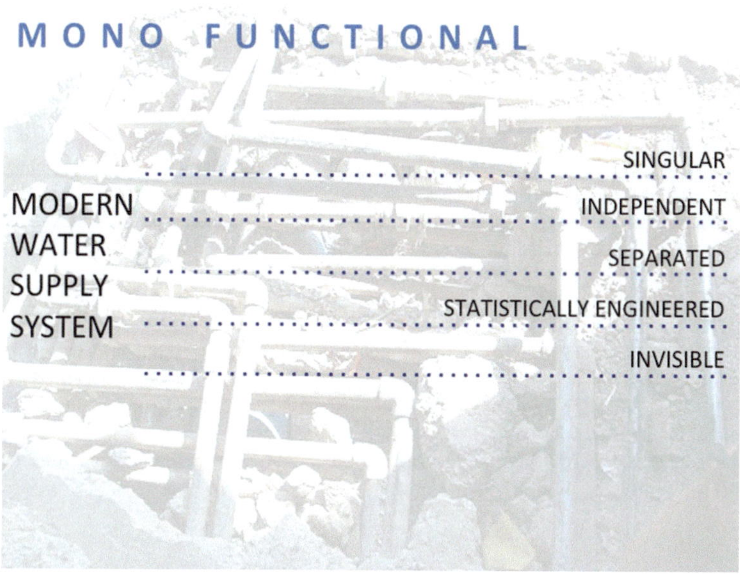

Fig. 8.6 The present-day water supply system has a singularity. Away from human interaction, it has taken away the connection of water and people

times have seen increasing awareness of environmental aspects in the infrastructure projects, they continue to be weighed in numerical and quantitative terms, which are simpler sometimes, to process and provide. The qualitative aspects, the people-nature relationship, and the various histories of the place are missing from the various discourses, policies, and reports (Fig. 8.6).

To balance the homogenizing and monocultural trends of the contemporary infrastructure developments, the case study of Jodhpur water bodies provides some clues. Its architecture incorporates the diverse components of urban living, offering a spatial experience that assimilates the network. The identification of the issues and problems becomes the first step toward an ecological urban regeneration process. It is hoped to initiate a productive dialogue on the integration of social and environmental ecosystems in infrastructure developments leading to urban regeneration. The aim is to redeem the ancestral and fundamental bond between the city and its waters.

All illustrations are by the author

References

Agarwal A (1995) Dismantling the divide between indigenous and scientific knowledge. Dev Chang 26:413–439

Agarwal A, Narain S (eds) (1997) Dying wisdom. Centre for Science and Environment, New Delhi

Berkes F (2012) Sacred ecology, 3rd edn. Routledge, New York

Coomaraswamy AK (1909) The Indian craftsman. Reprinted 2004. Munshiram Manoharlal Publishers Pvt. Ltd., New Delhi

Ground Water Board www.cgwb.gov.in. 01 07 2008. Government of India. [online] Available at http://www.cgwb.gov.in/District_Profile/Rajasthan/Jodhpur.pdf. Accessed 10 Jul 2014

Hegewald JAB (2002) Water architecture of South Asia. Koninklyke Brill, Leiden

Illich I (1985) H2O and the waters of forgetfulness. Dallas Institute of Humanities and Culture, Dallas

Iyer RR (2007) Towards water wisdom limits justice, harmony. SAGE Publications, New Delhi

Jain SK (2005) Dynamism of traditional knowledge. Indian J Tradit Knowl 4(2):115–117

Mishra A (2001) Radiant raindrops of Rajasthan (trans: Jani MG). Research Foundation for Science, Technology and Ecology (RFSTE), New Delhi. Print

Skyes PR, Huges (eds) (2000) Urban regeneration: a handbook. Sage Publications Ltd, London/ New Delhi. Print

Chapter 9
Assessing Impact of Sea Level Rise Along the Coastline of Mumbai City Using Geographic Information System

Pratibha D. Singh and A.R. Kambekar

Abstract A global mean sea level (MSL) rise of 1 m is being projected till the end of year 2100 by the Intergovernmental Panel on Climate Change (IPCC). Sea level rise (SLR) may directly affect the coastal zone in and around the world and may jeopardize the historic, cultural, and infrastructure resources. India, having a long coastline of around 7517 km (ISRO (2012) Coastal zone of India 2012. Ahmedabad: Space Applications Centre ISRO), is facing many problems due to discharge of waste effluents, municipal sewage, squatter settlements, exploitation of coastal resources, and continued development near the coast. The present study aims at developing SLR scenario using geographic information system (GIS) and its effect on the coastline of Mumbai city, Maharashtra, India. Single-value water surface method is used in ArcGIS 10.1 software to estimate the probable flooding for 1 m, 2 m, and 3 m sea level rise scenario. Comparative study for vulnerability intensity with respect to the probable area inundated and respective inhabitants affected for all the administrative wards coming under the Municipal Corporation of Greater Mumbai (MCGM) is prepared. This study shows that the level of loss of land due to coastal inundation might reach 3.84 %, 8.79 %, and 14.16 % for probable sea level rise of 1 m, 2 m, and 3 m.

Keywords Sea level rise • Infrastructure • Vulnerability • Sustainable city • GIS and Mumbai

9.1 Introduction

Sustainable city is the city which delivers economic, social, and environmental needs to the community without threatening the natural, built, and social system of community on which its services depends (Local 1994). The environmental factors

P.D. Singh (✉) • A.R. Kambekar
Sardar Patel College of Engineering, Mumbai, India
e-mail: singh.priti69@gmail.com; arkambekar@gmail.com

F. Seta et al. (eds.), *Understanding Built Environment*, Springer Transactions in Civil and Environmental Engineering, DOI 10.1007/978-981-10-2138-1_9

are important supports in building a sustainable structure in the new cities to be created, and it acts as a catalyst for the coexistence of environment and living habitants. As per IPCC fifth assessment report, the global sea level rise by the year 2100 will be from 0.28 m to 0.98 m under Representative Concentration Pathways (RCP) scenario (Wong et al. 2014). This will highly affect the coastal system and low-lying areas in the form of water submergence, coastal flooding, salt water intrusion, coastal erosion, and the corresponding population and assets exposed to it.

India, having a long coastline of around 7517 km being surrounded by Bay of Bengal in east, Arabian Sea to the west, and in the south by Indian Ocean (ISRO 2012), has a total coastal state and union territories population of around 540 million, whereas about 14.2 % of India's population leaves in coastal district of India (Institute for Ocean Management 2015), and large population of these reside in cities like Mumbai and Kolkata which are situated on the low-lying areas. Any sharp rise in the sea level could have a considerable impact on Indian geography given its long coastline and important cities located on the coast. Indian coastline has been observed to be inundated up to 34.906 km^2 due to rise in sea level till the twentieth century. The coastline in Maharashtra along Mumbai city has experienced a shift of 4.049 m and inundates an area of 16 km^2 (Nayak 2013). Both urban and rural areas being vulnerable to climate change, its impact may vary with the area having high concentration of population depending upon their geographical location and climatic conditions. The large infrastructure investments such as coastal roads, ports, industries, tourism, and other facilities in the coastal areas would be at greater risk due to rising sea levels in the future. Taking into account considerations of climate change, urban planners and policy-makers can address some of the issues that will assist in improvement of living conditions of the society. Cities in future need to be built to withstand the impact of climate change socially as well as structurally. Cities should therefore include mitigation and adaptation actions to take care of likely climate impacts, in conjunction with increasing the capacities of the populace, infrastructure, institutions, and governance (TERI 2014). It is very important to understand the level of vulnerability at a critical place before preparing mitigation and adaptation plan for a city. The aim of this paper is to assess the impact of sea level rise on the coastline of city of Mumbai, India, and also to find the vulnerability of each administrative wards or critical locations to provide solutions and to plan activities needed for disaster mitigation measures in advance to overcome effect of climate change in future.

9.2 Vulnerability of Mumbai

Mumbai being situated at a height of 10–15 m above the mean sea level (MSL) and at some places being at just above the sea level faces a major challenge from cyclones and flooding situations. Near about 136 areas of Mumbai having temporary houses exist within 500 m from the coast (Ministry of Environment & Forest 2010). Many fishermen's colonies are settled near the coastlines which are highly vulnerable. Erosion of coastline has been observed at Versova, Mumbai (Sanil Kumar 2006). Along with risk of sea level rise, Mumbai can face the same problem to a higher extent as of the year 2005, flooding in the year 2080 due to climate change (Hallegatte 2010). The city has experienced flooding due to heavy rainfall and high tide situation at the same time with extreme loss of life and assets in the month of July of the year 2005. The total loss occurred in damage of housing was 3.06 billion rupees (Government of Maharashtra 2005). The major source of above damage was the age-old drainage system of Mumbai as pointed out by the fact finding committee. While planning the city for disaster management, it is very much important to consider the hazard to which the city is exposed and also the impact of climate change on it (Ranger 2011).

9.3 Study Area

The study area is Greater Mumbai region of Maharashtra, India, situated between 18 and 19.20° northern latitude and between 72 and 73.00° eastern longitude having a long coastline of 167 km (Maharashtra Emergency Earthquake Rehabilitation 2005); the area is surrounded on three sides by sea: by the Arabian Sea to the west, the south by the Harbour Bay, and the Thane Creek on the east. Greater Mumbai is divided in three parts: the island city, the eastern suburbs, and western suburbs having 24 administrative wards (A, B, C, D, E, F/N, F/S, G/N, G/S, H/W, H/E, K/E, K/W, L, M/E, M/W, N, P/N, P/S, R/N, R/C, R/S, S, T) as shown in Fig. 9.1 in total covering an area of 458.28 km^2 including rocky outcrops, beaches, BARC area, and other land masses (MCGM 2014).

9.3.1 Approach

Esri ArcGIS 10.1 software is used for data preparation and data analysis purpose. The data used for analysis work is given in Table 9.1. Mean sea level data for Mumbai region is obtained from the Global Sea Level Observing System (GLOSS). A digital elevation model (DEM) is derived using NASA Shuttle Radar Topography Mission (SRTM) with 90 m resolution (Jarvis 2008). The ward boundary of Mumbai is defined using the ward plan from Municipal Corporation of Greater Mumbai

Table 9.1 Data used

Sr. no.	Parameters	Data used	Resolution	Period
1.	Sea level change rate	GLOSS data	–	1900–2010
2.	Elevation data	SRTM	90 m	–
3.	Ward map of Mumbai	MCGM	–	1991
4.	Socioeconomic data	MCGM	–	2011

Source: Author

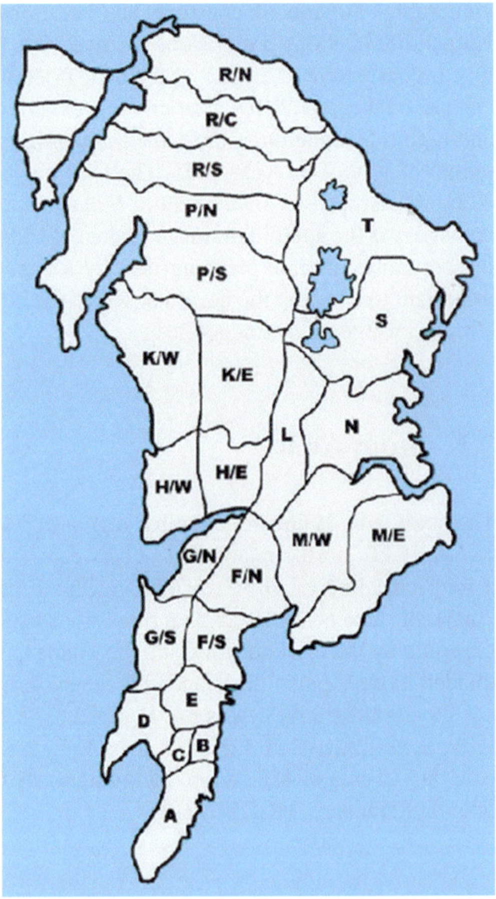

Fig. 9.1 Administrative ward map of Greater Mumbai (Source: http://www.mcgm.gov.in/)

(MCGM 2014). Population and area distribution data is taken from the sanctioned development plan of Mumbai (MCGM 2014).

9.4 Methodology Adopted

For mapping the area inundated under the sea level rise scenario, the single-value water surface method (fill the bath tub method) ((NOAA) Coastal Services Center 2012) is used with ArcGIS 10.1 software. In this method with help of data obtained from GLOSS of Apollo Bunder (Station ID 43), the mean sea level of Mumbai was determined. To get the area of land inundated, administrative ward-wise map of Mumbai was geo-referenced and digitalized, and it was divided into 24 administrative wards as per municipal cooperation of Mumbai. In this study

M/E and M/W are considered as ward M, likewise P/N and P/S are considered P ward and R/N, R/C, and R/S as R ward. For each coastal ward, the area inundated was calculated with the help of DEM. The SRTM data of 90 m resolution was used to derive digital elevation model (DEM). Raster Calculator tool from spatial analyst tool box was used for calculating the area inundated under the sea due to climate change in 2100 for rise of 1 m, 2 m, and 3 m above mean sea level, respectively. The obtained result from the Raster Calculator was reclassified to get only the area inundated using the reclassify tool from spatial analyst tool; the reclassified map was converted to vector form for calculating the area. The vector map of the wards was then projected to WGS 1984/UTM Zone 43 N to get the area inundated in required units. Along with the hydrologically connected area, the low-lying area is also considered in inundation map as the low-lying area will get highly affected in the climatic event of both high tide and heavy rainfall as that of condition similar to 2005 Mumbai flood. Geometrical increase method was used to project the population (United Nations 1952) of the year 2101, and the predicted population likely to be affected under various SLR scenarios is estimated. In this method the population (P_n) at end of tth decade can be estimated by the following formula:

$$P_n = P_o(1+r)^t \tag{9.1}$$

where:

P_o is population at beginning of period.
r is the annual rate of increase.
t is the period of time in years.
P_n is the population at end of period.

The growth rate of population of all administrative wards of Mumbai has been taken as the cumulative growth rate of Greater Mumbai region. The population data of the year 2001 and 2011 is used to determine the value of "r" in Eq. 1.

After predicting the population of 2101 based upon the baseline year of 2011, the population affected for a particular area inundated in each administrative ward is calculated as follows:

$$P_a = A_i * P_d$$

where:

P_a is population affected in a ward.
A_i is the area inundated under different sea level rise scenarios (sq.km).
P_d is the population density of a ward in 2101 (person/sq.km).

9.5 Result and Discussion

It has been observed from the analysis that the coastline of the city of Mumbai will be vulnerable to climate change in the future due to possible sea level rise. Figures 9.6, 9.7, and 9.8 show the possible inundation as seen in red color due to 1 m, 2 m, and 3 m sea level rise. A 1 m sea level rise may affect approximately 3.841 % of land area including the natural marshy areas and residential areas with possible relocation of a population of around two million (3 %) of projected population in 2101. Under 2 and 3 m scenarios along with the marshy land, the residential area in F/N, M, N, S, T, P, and R wards will face inundation problem both temporarily and

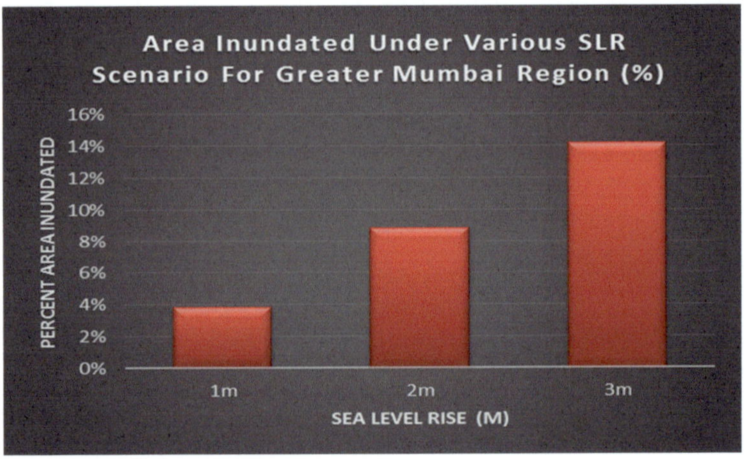

Fig. 9.2 Area inundated under various SLR scenarios (Source: Author)

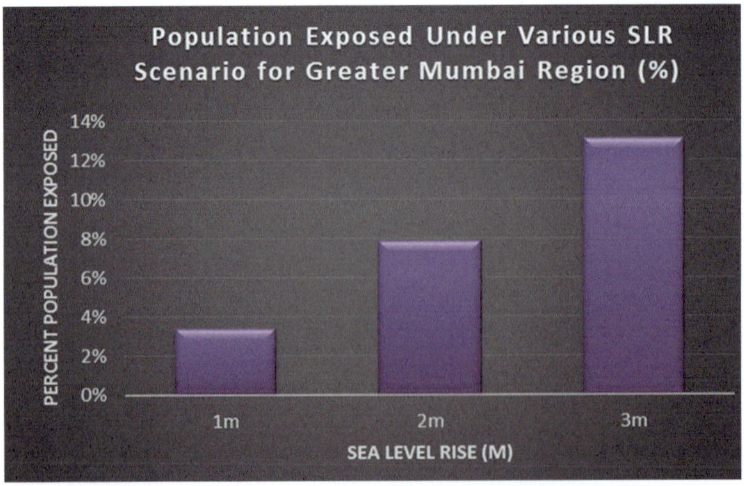

Fig. 9.3 Population exposed under various SLR scenarios (Source: Author)

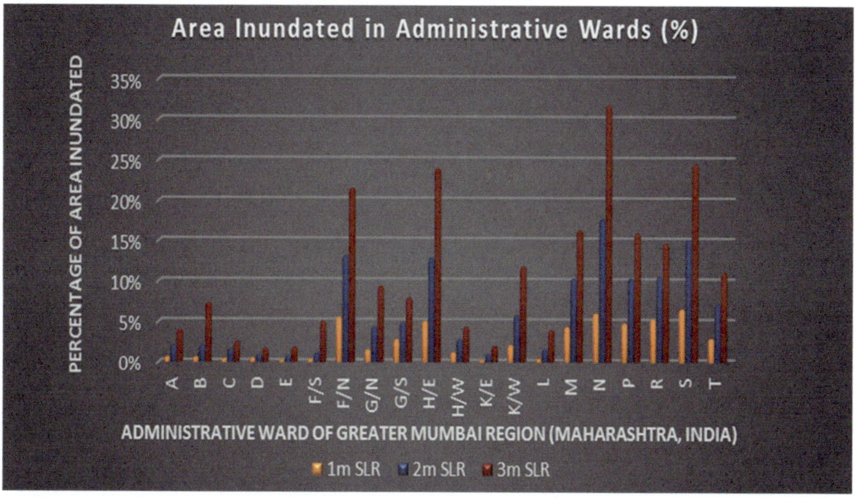

Fig. 9.4 Ward-wise area inundated in year 2101 under 1 m, 2 m, and 3 m SLR scenario (Source: Author)

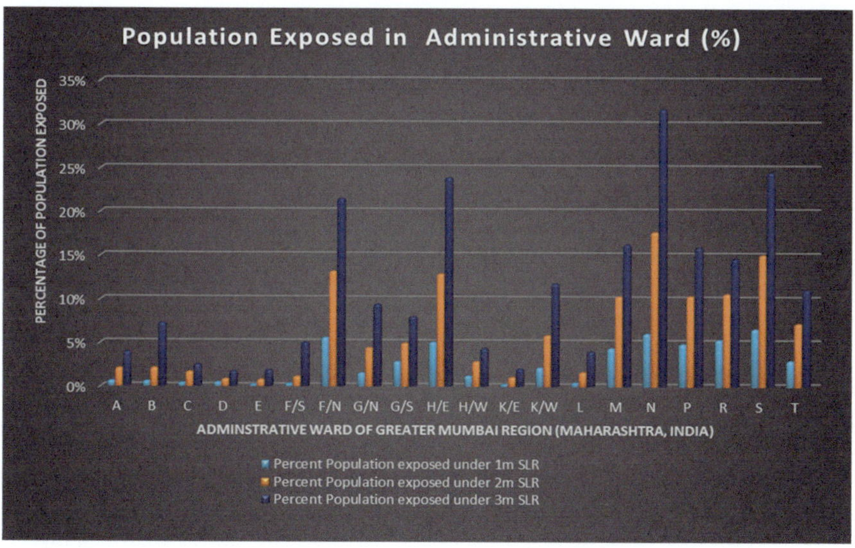

Fig. 9.5 Ward-wise population expose in year 2101 under 1 m, 2 m, and 3 m SLR scenario (Source: Author)

permanently. The industrial area of administrative ward M/W will face challenges due to increase of SLR by 3 m. The increase in sea level rise will also bring a concern to the transportation area of administrative ward "B." Major devastating condition might be possible under 3 m SLR affecting around 14.16 % of the total area and a population of 7.8 million (13 %) of total projected population (refer to Figs. 9.2 and 9.3). The areas that may be affected critically due to different SLR scenarios are

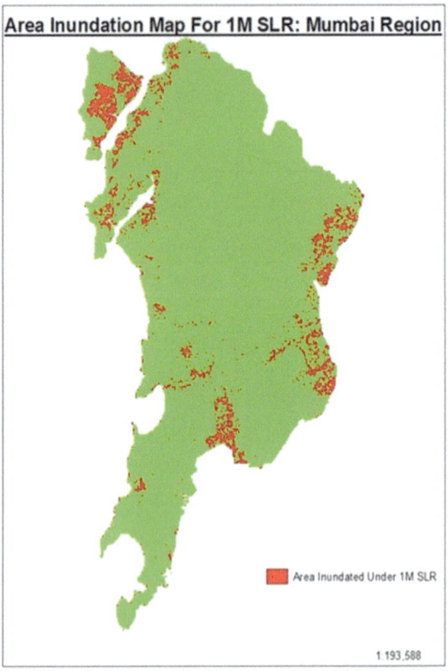

Fig. 9.6 Area Inundation MaP for 1m SLR (Source: Author)

M, N, P, R, S, and T wards of Mumbai. Possible inundation in R ward of Mumbai may be around 4.34 km^2 (5 %) of area of the ward for 1 m SLR; the area inundated will be the marshy lands and open areas along the coast and around 11.84 km^2 (16 %) of area for 3 m SLR which possibly will include the residential area of that ward affecting a population of about 5.3 % and 14 % of projected population of that ward in the year 2101, respectively, as shown Figs. 9.4 and 9.5. As mangrove acts as a natural buffer to Mumbai city, it will be affected critically by the 1–3 m SLR.

The major population situated near the coastline of Mumbai would be affected due to the above scenario. About 3.84 %, 8.79 %, and 14.16 % of total land would possibly inundate under probable sea level rise of 1 m, 2 m, and 3 m affecting a population of about 3 %, 8 %, and 13 % of projected population for year 2101 as shown in Figs. 9.2 and 9.3. As being made up of reclaimed land, there are chances of subsidence of land at many places in coming years, and those areas need to be critically assessed for vulnerability due to sea level rise. The input is taken from the SRTM 90 m resolution being a coarse data; it has some limitations. It is recommended that assessment of future inundation of coastline and cities due to possible SLR shall be done with fine resolution data. As sea level rise is a global phenomenon, many countries situated near the coastline have already started planning for mitigation and adaptation measures to confront the future disaster situation. It is definitely possible to deal with the impact on civil engineering projects along the coastline. Urban planning of coastal cities shall include spatial planning measure to take care of any kind of coastal hazards with respect to climate change.

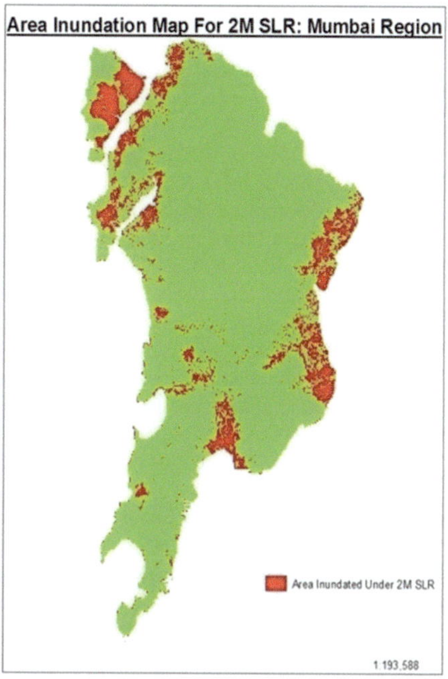

Fig. 9.7 Area Inundation MaP for 2m SLR (Source: Author)

Fig. 9.8 Area Inundation MaP for 3m SLR (Source: Author)

References

Government of Maharashtra (2005) Maharashtra floods 2005: relief and rehabilitation. Status report. Mumbai

Hallegatte (2010) Flood risks, climate change impacts and adaptation benefits in Mumbai: an initial assessment of socio-economic consequences of present and climate change induced flood risks and of possible adaptation options. OECD Environment Working Papers

Institute For Ocean Management. Database on coastal states of India. From Centre For Coastal Zone Management And Coastal Shelter Belt: http://iomenvis.nic.in/. Retrieved 13 Mar 2015

ISRO (2012) Coastal zone of India 2012. Space Application Cetre ISRO, Ahmedabad

Jarvis A (2008) Hole-filled seamless SRTM Data V4. Retrieved From International Centre For Tropical Agriculture (CIAT): http://srtm.csi.cgiar.org

L. A (1994) International Council For Local Environmental Initiatives (1994) Local Agenda 21 Participants Handbook. ICLEI, Toronto.

Maharashtra Emergency Earthquake Rehabilitation (2005) Greater Mumbai disaster management action plan risk assesment & response plan Vol I & II. Mumbai. Municipal Co-Operation Of Greater Mumbai

MCGM Existing land use maps and report. From http://www.mcgm.gov.in. Retrieved Nov 2014

Ministry of Environment & Forest (2010) Proposed coastal regulation zone (CRZ) amendments, 2010. Government of India, India

Nayak S (2013) Estimation of Indian coastal areas inundated into the sea due to sea-level rise during the 20th century. Curr Sci 104(5):583–585

NOAA-Coastal Services Center (2012) Mapping coastal inundation primer. National Oceanic and Atmospheric Administration

Ranger N (2011) An assessment of the potential impact of climate change on flood risk in Mumbai. Springer Clim Chang 104:139–167

Sanil Kumar V (2006) Coastal processes along the Indian coastline. Curr Sci 91(4):530–536

TERI (2014) Climate proofing Indian Cities: a policy perspective. Teri Policy Brief (March 2014)

United Nations (1952) Manual I: methods for estimating total population for current dates. United Nations Publications, Sales No. 52.XIII.5

Wong et al (2014) Coastal systems and low-lying areas. In: Climate change 2014: impacts, adaptation, and vulnerability. Part A global and sectoral aspects. UK: Cambridge University Press

Chapter 10
Sustainable Landscaping in Cyclone-Prone Areas: A Paradigm Shift

K. Mohan and Jagadeesh Gorle

Abstract Cyclone Hudhud, which struck Visakhapatnam on October 12, 2014, delivered a crippling blow to the green cover of the city. Most of the trees in public parks, university campuses and residential layouts in the vicinity of the Beach Road suffered from extensive or partial damage. Majority of the flowering and foliage trees suffered from discoloration of leaves due to salt spray deposited by the cyclonic winds. A post-cyclone damage assessment to the vegetation along the coastal corridor has been carried out by the authors with the objective of understanding the vulnerability of different species of trees to cyclonic storms. Several indigenous species survived the fury of the cyclone, whereas many exotic species succumbed. The present study has identified various types of damage sustained by trees under the influence of strong cyclonic winds. Post-cyclone, some of the beaches in Visakhapatnam have also suffered from extensive erosion. The significance of preserving the indigenous groundcovers for arresting beach erosion is also discussed. Plant selection criteria for landscaping normally include an assessment of their visual, ornamental and functional characteristics along with their cost, availability, etc. However in cyclone-prone areas, a paradigm shift in plant selection criteria is essential. Trees and shrubs need to be selected that are indigenous, wind resistant, possess a minimum of co-dominant stems and are resistant to damage from salt spray. This would help in minimising the damage to trees from future storms and help in the creation of safe and sustainable coastal environments.

Keywords Sustainable landscaping • Indigenous vegetation • Salt spray effect • Wind-resistant trees • Shelter belts

---

K. Mohan (✉) • J. Gorle
Gitam School of Architecture, Gitam University, Visakhapatnam, India
e-mail: profkmohan@gmail.com; architectjagadeesh@gmail.com

F. Seta et al. (eds.), *Understanding Built Environment*, Springer Transactions in Civil and Environmental Engineering, DOI 10.1007/978-981-10-2138-1_10

10.1 Introduction

Most of the green cover along the coastal corridor of Visakhapatnam suffered from extensive damage in the wake of the very severe Cyclone Hudhud. A post-cyclone damage assessment to the vegetation along the coastal corridor has been carried out by the authors with the objective of understanding the vulnerability of different species of trees. The study has brought out the need for a paradigm shift in landscaping thought for cyclone-prone areas. The approach to landscaping in cyclone-prone areas presents a challenge to the designer. Plants need to be carefully selected after assessing their tolerance for strong winds and salt in the air. Trees need to be assessed for their structure and rooting pattern. As quick growing trees tend to have shallow roots, they are highly susceptible to damage from storms. An inspection of trees before and after a storm is essential for landscaping to be sustainable in cyclone-prone areas.

10.2 Climate of Visakhapatnam

Visakhapatnam is at $17^0 42'$ N latitude, 83^0 $17'$ E longitude and 4 m (11 ft) above mean sea level. It has a tropical wet and dry/savanna climate with a pronounced dry season in low-sun months and no cold season. The wet season is in the high-sun months. The annual average temperature is 28.4 °C, and average monthly temperatures vary by 8.4 °C. May is the hottest month having an average temperature of 32.1 °C. The total annual precipitation averages 955 mm (37.6. inches). The average annual relative humidity is 71.8 % and average monthly humidity ranges from 67 % in December to 78.2 in June. Annual average wind speed is 5.39 m/s, varying from 4.55 m/s in January to 6.38 in July (source: Climatemps.com).

Due to its proximity to the Bay of Bengal, Visakhapatnam is mostly prone to depressions that arise from the sea and is also occasionally vulnerable to tropical storms. From 1891 to 2014 as per the report on Cyclone Hazard Prone Districts of India (Mandal and Mahapatra 2010; Kamalakar 2014), Visakhapatnam is the first Indian city that has suffered a direct hit by a high-intensity cyclone. Hudhud made its landfall at Kailasagiri, Visakhapatnam, on October 12, 2014. It battered the coast with wind speeds exceeding 180 kmph and caused extensive damage to infrastructure, buildings and the city's green cover.

10.3 Various Types of Damage to Trees

Trees can get damaged in various ways during a storm. The damage can be partial such as defoliation, breakage of stems or severe as in the case of uprooting or damage from salt spray deposited on leaves by cyclonic winds.

Fig. 10.1 (**a**) A defoliated *Ficus religiosa* near Rushikonda after the cyclone Hudhud. (**b**) *Ficus religiosa* tree with new sprouts 15 days after the cyclone

Fig. 10.2 (**a**) A defoliated *Terminalia catappa* tree after the cyclone Hudhud. (**b**) *Terminalia catappa* tree with new sprouts 2 weeks after the cyclone

10.3.1 Damage to Canopy or Defoliation

Defoliation or loss of leaves from the canopy in the face of strong cyclonic winds is a defensive strategy adopted by many trees for reducing the load on the stems. This type of damage many a time is not very severe, and trees normally sprout new leaves after the cyclone (Kamph et al. 2007). Photos of a defoliated and regenerated *Ficus religiosa* tree along the Beach Road near Rushikonda are shown in Fig. 10.1a and 1, respectively. Figure 10.2a shows a defoliated *Terminalia catappa* tree which lost almost its entire foliage during the cyclone. Figure 10.2b shows the same tree which regained its entire foliage 2 weeks after the cyclone. Trees with large leaves which do not shed easily are highly vulnerable in the face of strong winds. Figure 10.3a, b show a broad-leaved *Ficus auriculata* tree in the Chandana Mohan Rao Nursery at Visakhapatnam before and after the cyclone, respectively.

Fig. 10.3 (**a**) View of *Ficus auriculata* tree in the Chandana Mohan Rao Nursery at Visakhapatnam before the cyclone. (**b**) *Ficus auriculata* tree damaged by cyclone Hudhud

Fig. 10.4 (**a**) Uprooted *Swietenia mahagoni* trees at the Gitam University campus. (**b**) An uprooted *Spathodea campanulata* tree at Gitam University campus

10.3.2 Wind Throw and Uprooting of Trees

Tall, slender and top heavy trees with shallow roots are vulnerable to wind throw or uprooting during a storm. Figure 10.4a shows uprooted *Swietenia mahagoni* trees (exotic top heavy species with shallow roots), being restored in the Gitam University campus at Visakhapatnam. Figure 10.4b shows an uprooted *Spathodea campanulata* tree (another exotic variety) in front of the Architecture School at Gitam University. On several urban roads in Visakhapatnam, many *Peltophorum* trees either suffered from broken branches or were completely uprooted during the Hudhud cyclone. Insufficient space for the roots of trees planted adjacent to roads in the city of Visakhapatnam ultimately resulted in the uprooting of a large number of *Peltophorum* and *Ficus religiosa* trees.

Cyclone Hudhud caused extensive damage to the trees in the Andhra University campus at Visakhapatnam. Most of the *Eucalyptus* trees planted close to buildings

(a) (b)

Fig. 10.5 (**a**, **b**) Uprooted *Eucalyptus* trees in the Andhra University campus at Visakhapatnam

(a) **(b)**

Fig. 10.6 (**a**) A wind thrown *Acacia auriculiformis* tree on the Beach Road and standing *Pongamia pinnata* trees. (**b**) A *Peltophorum* tree with split branches atop the Kailasagiri hill where Hudhud made its landfall

got uprooted, and several buildings bore the brunt of the falling trees. Figure 10.5a, b show uprooted and damaged *Eucalyptus* trees in the aftermath of the cyclone in the Andhra University campus.

Many young *Pongamia pinnata* trees planted along the Beach Road survived without much damage. However most of the *Acacia auriculiformis* trees whose bark is inherently brittle suffered from broken stems and many of the Acacia trees also got uprooted. Figure 10.6a shows the uprooted *Acacia auriculiformis* tree between the standing *Pongamia pinnata* trees along the Beach Road. Figure 10.6b shows a view of the Beach Road as seen from the Kailasha Hill where Hudhud made its landfall. Also seen in the picture is a storm-damaged *Peltophorum* tree with broken stems. Many *Peltophorum* trees were uprooted and damaged on Kailasha Hill as well as along the Beach Road. Figure 10.7a shows one such uprooted tree. Figure 10.7b shows an uprooted *Peltophorum* tree in Shivaji Park at Visakhapatnam.

(a) (b)

Fig. 10.7 (**a**) Uprooted *Peltophorum* tree along the Beach Road near Rushikonda, Visakhapatnam. (**b**) An uprooted and damaged *Peltophorum* tree in Shivaji Park, Visakhapatnam

Fig. 10.8 Damaged co-dominant stems of *Ficus religiosa* showing bark inclusion

10.3.3 Breakage of Stem and Branches

Breakage of stems and branches during strong winds is a phenomenon that has been observed in trees with co-dominant stems. Co-dominant trees have two or more stems of similar diameter. A tree with a single dominant stem is much stronger than a tree with multiple co-dominant stems. Inferior wood or inclusion bark develops between the V-shaped joints in co-dominant stems, making them susceptible to breakage during storms. Tests on co-dominant stems at the Bartlett Tree Research Laboratories in Charlotte, North Carolina, have brought out the weakness of co-dominant stems with inclusion bark (Kane and Clauston 2008; Smiley 2003). The broken co-dominant stem of a *Ficus religiosa* tree along the Beach Road due to strong winds during Hudhud is shown in Fig. 10.8. A *Pongamia pinnata* tree with broken co-dominant stems in the Gitam University campus is shown in Fig. 10.9.

Fig. 10.9 Damaged co-dominant stems of *Pongamia pinnata* in the Gitam University Campus at Visakhapatnam

Fig. 10.10 (**a**) Salt spray damage to a leaf of Ficus elastica at Tenneti Park, Visakhapatnam. (**b**) and (**c**) salt burned leaves of *Bauhinia blakeana* at Tenneti Park, Visakhapatnam

10.3.4 Damage by Salt Spray

Seaside plants most frequently suffer from salt spray damage. Strong winds deposit water droplets laden with salt on trees and shrubs; this prevents the plant from absorbing water properly and leads to plant dehydration. Curling of edges, reduction in size of roots, stem and leaves and ultimately death may result from salt spray damage (Sinclair et al. 1987; Seigendorf 1984). Salt spray damage to the leaves of Ficus elastica and *Bauhinia blakeana* trees in the Tenneti Park, located along the Beach Road at Visakhapatnam, are shown in Fig. 10.10a–c, respectively.

Salt spray damage to the leaves of *Calophyllum inophyllum* trees and *Tecoma grandis* trees in the Gitam University campus is shown in Fig. 10.11a. Salt-burned *Mimusops elengi* trees in the Gitam University campus at Visakhapatnam is shown in Fig. 10.11b. Figure 10.11c shows a *Bauhinia blakeana* tree with leaves damaged by salt spray in the Vuda Park at Visakhapatnam. *Nerium oleander* shrubs, planted

(a) (b) (c)

Fig. 10.11 (**a**) Salt spray damaged *Calophyllum inophyllum* trees (*left*) and *Tecoma grandis* trees (*right*) at the Gitam University campus, Visakhapatnam. (**b**) Salt-burned *Mimusops elengi* trees in the Gitam University campus at Visakhapatnam. (**c**) Salt-burned leaves of *Bauhinia blakeana* at Vuda Park, Visakhapatnam

Fig. 10.12 (**a**) *Nerium oleander* shrubs. (**b**) *Cassia biflora* and *Borassus flabellifer* palms

along the medians on the Beach Road, have shown remarkable tolerance to salt spray and have also proved to be highly wind resistant (Fig. 10.12a). Figure 10.12b shows *Cassia biflora* plants and *Borassus flabellifer* palms which withstood the onslaught of the cyclone without any damage.

Table 10.1 summarises the various types of damage sustained by different trees and palms observed from the damage assessment study.

10.3.5 Beach Erosion, Shelter Belts and Soil Stabilisers

Post-Hudhud, extensive erosion has been noticed at several stretches along the Visakha-Bheemili coastal corridor, especially in front of Kirlampudi Layout and Ramakrishna Beach. The beach wall has been breached at several places at Kirlampudi Layout (Fig. 10.13b).A survey of the *Casuarina equisetifolia* shelter

Table 10.1 Type of damage sustained by trees and palms

Scientific name of tree	Type of damage/comments
Terminalia catappa	Good tree architecture, wind and salt resistant
Mimusops elengi	Wind sturdy but leaves prone to salt spray damage
Spathodea campanulata	Shallow roots, easily uprooted, co-dominant stems
Polyalthia longifolia	Stable structure but leaves prone to salt spray damage
Azadirachta indica	Small leaves, quick recovery of foliage after storm
Tectona grandis	Large leaves which do not fall, easily uprooted
Tabebuia argentea	Brittle branches that break easily
Dalbergia sissoo	Brittle branches, not wind resistant
Eucalyptus spp.	Uprooted mature trees can damage buildings
Swietenia mahagoni	Top heavy foliage and shallow roots. Not wind resistant
Delonix regia	Brittle branches which tend to break easily in a storm
Pongamia pinnata	Younger trees stable, mature braches not wind resistant
Bauhinia blakeana	Co-dominant stems. Leaves affected by salt spray
Acacia auriculiformis	Brittle branches that break easily and trees gets uprooted
Calophyllum inophyllum	Single stem, handsome foliage but leaves not salt resistant
Nerium oleander	Stable architecture. Wind and salt resistant
Cassia biflora	Stable architecture. Wind and salt resistant
Araucaria heterophylla	Pyramidal form highly wind resistant. Salt resistant
Caryota urens	Top heavy foliage. Easily gets uprooted in a storm
Pisonia alba	Co-dominant stems which break easily
Cocos nucifera	Top heavy. Fronds get damaged but are replaced gradually
Veitchia merrillii	Highly wind resistant and resistant to salt spray damage
Hyophorbe lagenicaulis	Aerodynamic form, resists wind and salt spray damage
Bismarckia nobilis	Big sized fronds, not wind resistant
Bambusa vulgaris	Stable stems in groups. Leaves salt resistant
Wodyetia bifurcata	Highly wind and salt resistant
Thespesia populnea	Highly salt resistant. Ideal for mangroves and shelterbelts
Alstonia scholaris	Good tree architecture, wind and salt resistant
Padanus	Native to coastal areas, wind and salt resistant
Ficus auriculata	Very big leaves which don't fall. Easily falls in storms
Casuarina equisetifolia	Good for shelter belts if densely planted. Stems brittle

belts along the Visakha-Bheemili coast has shown that shelter belts planted in dense rows survived the cyclone well, whereas thin rows of *Casuarina* tree sustained extensive damage (Kotamrazu et al. 2014; Kotamrazu 2014).

The study has identified several species of ground covers that have adapted to the coastal climate of Visakhapatnam. One such groundcover, *Ipomoea pes-caprae*, is found in several stretches of the beach near Kirlampudi Layout (Fig. 10.14a). This creeper acts as a soil stabiliser and is also salt tolerant. Another quick growing creeper, *Antigonon leptopus*, found extensively along the beach near Bheemunipatnam is shown in Fig. 10.14b. These creepers, if planted extensively along the beach, would help in further arresting erosion of the beaches in the long run.

Fig. 10.13 (a) *Casuarina* shelter belt that survived the onslaught of Hudhud. (b) The broken staircase leading to the badly eroded beach at Kirlampudi Layout, Visakhapatnam

Fig. 10.14 (a) *Ipomoea pes-caprae* growing on the beach at Kirlampudi Layout, Visakhapatnam. (b) *Antigonon leptopus* growing on the beach near Bheemunipatnam

10.4 Conclusions

For landscaping to be sustainable in coastal areas, the plant species should be wind resistant, salt tolerant and sufficiently resilient to recover in the aftermath of a storm. Damage to trees can be minimised by selecting trees with single dominant stems rather than trees with multiple co-dominant stems. Trees in public parks and in localities in the vicinity of the coast should be inspected for post-cyclone damage, and a structural pruning programme should be put in place. Native and indigenous trees should get preference over exotic species during plant selection as they adapt well to local conditions. Plantations of dense mangroves and shelter belts can help as a first defence from the onslaught of cyclonic storms. Indigenous ground covers should be planted for dune stabilisation and reduction of beach erosion. Such measures would help in the reduction of damage to landscape elements from cyclonic storms in future.

References

Kamlakar R (2014) Vizag first Indian City directly hit by Cyclone Hudhud. http://timesofindia.indiatimes.com/city/visakhapatnam/Vizag-first-Indian-city-directly-hit-by-cyclone-Hudhud/articleshow/44864271.cms

Kampf E, Duryea ML, Gilman EF, Delgado A (2007) Assessing hurricane-damaged trees and deciding what to do. Available at http://www.leegov.com/Documents/

Kane B, Clouston P (2008) Tree pulling tests of large shade trees in the genus Acer. Arboricult Urban For 34(2):101–109

Kotamrazu M (2014) Wind damage to trees in the Gitam University Campus at Visakhapatnam by Cyclone Hudhud. IJRET 3(Special Issue 16):55–65

Kotamrazu M, Gairola A, Kwatra N (2014) Wind damage to trees at Visakhapatnam by Cyclone Hudud. Proceedings: 7th National Conference on Wind Engineering (NCWE-2014), Thapar University Patiala, pp 272–287

Mandal GS, Mohapatra M (2010) Cyclone hazard prone districts of India: a report. www.ndma.gov.in/images/cyclones/cyclonepronedistrict.pdf

Seigendorf L (1984) Hurricane tree care. J Arboric 10:217–221

Sinclair WA, Lyon HH, Johnson WT (1987) Diseases of trees and shrubs. Cornell University Press, Ithaca, p 575

Smiley ET (2003) Does included bark reduce the strength of codominant stems? J Arboric 29(2):104–106

www.visakhapatnam.climateps.cmom

Chapter 11
Urban Regeneration and Social Sustainability of Indore City

Soma Anil Mishra, R.K. Pandit, and Mayank Saxena

Abstract Sustainability is the most frequently referred term among urban thinkers. Rapid urbanization has compelled us to think about urban sustainability. Various factors affect urban environmental conditions. The idea of sustainable development lays emphasis on encompassing all the three parameters of sustainability. There must be balance between socioeconomic activities and environment in an urban setting. Process of development should enhance the quality of human life. Indore city has a population of more than two million. Over the period of time, Indore has been regenerated dynamically. Regeneration has affected the sustainability of Indore city. The paper lays stress on social sustainability, focusing on the importance of safe society for a livable community and sustainable development. It is the responsibility of architects and urban planners to aim for creating safe environment which is attractive to citizens and rich in ecological sense. Research paper deals with the above aspect and inspect of Indore city through a process of dividing it into four major sections based on demographics and crime rate. The researcher seeks to evaluate the relationship of urban regeneration (UR), social sustainability, and environmental design of Indore city. The research makes use of obtainable sources and information from previous findings and research. The primary data is obtained as information from questionnaire. Then, the data is subjected to ANOVA and Amos software of SPSS 21 and linear regression analysis. Finally the paper proves that both urban regeneration and physical environmental design have considerable impact on social sustainability.

Keywords Urban • Urban regeneration • Social sustainability • Crime • Safe society • Environmental design

---

S.A. Mishra (✉) • R.K. Pandit
M.I.T.S, Gwalior, Gwalior, India
e-mail: somanil.mishra@gmail.com; drpanditrk@yahoo.com

M. Saxena
I.I.M.R, Indore, Indore, India
e-mail: mayank.saxena71@gmail.com

F. Seta et al. (eds.), *Understanding Built Environment*, Springer Transactions in Civil and Environmental Engineering, DOI 10.1007/978-981-10-2138-1_11

11.1 Introduction

City is a living organism. It is one of the greatest human creations on earth. Any achievement of the past may become obsolete in future context, hence regeneration of old and generation of new ideas are essential for any living organism like a city to survive and grow. In the dynamic urban system, regeneration is inevitable. The challenge is how a city growth maintains balance with sustainability. Social sustainability is essential for equitable and livable community. Urban designers need to realize their duty of contribution for safe and sustainable community. Physical environment can create safe or unsafe society.

11.2 Definition of Urban

To show more power and hierarchy, urban areas were built with boundaries and distinguished structures (Wright 1953). An Urban Centre is based on the idea of an integral entity which encompasses a combination of activities which dominate urban life. According to Corbusier, the activities are living, working, cultivating the body and mind, and moving about. The Urban Centre occupies a privileged position in architecture design.

Urban is "the place of the unexpected." (Davis 1995). City is an entity; it is direct present reality and tangible architectural fact, but "urban" can be explained as a social and mental reality (Rossi 1984).

Hence Urban Centre is:

- An organism: It also has life having all the components of living organism. It needs fresh air to breathe, freshwater reservoirs for its occupants, comfortable vehicular and pedestrian movement, energy-efficient housing, and above all a livable atmosphere.
- Holistic hub of opportunities: An Urban Centre has opportunity for each and every citizen including job opportunities and educational resources. It is the main reason of rural migration. Sustainable ways of opportunities should be worked out with large life span so that nobody is jobless; this in turn can reduce crime rate of urban centers.
- Expression of cultures: People from all types of background gather here in search of opportunities carrying with them their culture; an Urban Centre expresses this cosmopolitan attitude and welcomes all its citizens.
- Accumulation of amenities: An Urban Centre while fulfilling the needs of its occupants is full of amenities with the job of expertise team members.
- Representation of social, economic, and environmental status of society: An Urban Centre represents the status of all its neighboring vicinity. It has the essence of the same in a polished manner.

11.3 Urban Regeneration

Urban regeneration is re-creation or revitalization of urban cells. Growing cells regenerate naturally; declining cells have to be regenerated. City is not a material morphological element; rather it is mental and social morphology. City is a social element and reflection of how citizens gather, converge, or diverge for their day-to-day activities. City is a spatial object, mediation, and a work (Lefebre 1996). Urban regeneration or regeneration of urban architecture and urban society is an essential prerequisite for societies in change and for the dynamic role of cities' contribution to national development. An urban system is a mutating organism, i.e., a living organism which is alive and transforming (Graham 2005). The city is the local origin of various environmental problems. Environment here is referred to atmospheric, sociocultural, as well as economic issues prevailing in urban setup. In the dynamic urban system, some areas grow and some areas decline; to meet both the needs, cells (areas) are regenerated to keep the system rejuvenating.

As per author's understanding, growth and decline support each other because household, business, and economy keep on moving without interruption from one location to another looking for "good life" and economic and social appreciation, hence supporting urban regeneration.

11.4 Social Sustainability

Sustainable society, the term coined in Coop America Quarterly, No. 37 Summer 1995, is a society which has a long-term goal for continuing to exist in good condition and reputation. It emphasizes on preserving the environment, also having good and peaceful relationships with neighboring societies. The precondition of this is access to:

- Basic supply (of food, clothing, habitation, and health)
- Independent security (security by government or voluntary activities)
- Equal opportunities (to education, information, and employment)
- Social resources (social cohesion, tolerance, solidarity, nonviolence)

Social sustainability demands every citizen's right to participate in his/her neighborhood activity as an essential element. Social sustainability can be achieved by positively affecting the built environment (Glasson and Cozens 2010). Better environments can be created where citizens do not have fear of crime and people are not indulged into crime (Newman 1972). A community is socially sustainable when it is safe, when the citizens perceive it to be safe, and when others also consider it to be safe (Cozens et al. 2005). Planning professionals have a main and pivotal role when an urban system desires to promote sustainable development.

11.5 Indore City

11.5.1 Historical Background

Situated on one of India's oldest pilgrimage routes from Mahakaal at Ujjain to Rameshwaram, Indore was a convenient resting place since historical times. It was on the route of the Marathas of Deccan on their way to North India. Trading communities from all over the region came in large numbers to settle in and around this new city. The city became the capital of Indore princely state in 1818. Between 1948 and 1956, Indore served as the summer capital of the former Madhya Bharat state established after the independence of India in 1947. Currently it is the commercial capital of state of Madhya Pradesh (John Malcom 1985).

11.5.2 Geography

Indore is located in the western region of Madhya Pradesh state in India, on the southern edge of the Malwa plateau. Indore has an average elevation of 550 m above sea level.

11.5.3 Population Growth and Urbanization of the City

Area of Indore city is 131.0 km^2 (source: City Development Plan 2021); population is 2,000,000. Population increase between 2001 and 2011 has been tremendous; this growth mainly attributes to the rapid industrial and commercial development leading to migration (Table 11.1). This population growth exerts pressure on natural resources and man-made infrastructure.

With this unprecedented growth, new challenges of social sustainability are heading high; especially citizens' safety is a matter of concern (Mishra and Pandit 2013).The average growth rate of Indore has been 40 % which is higher than the

Table 11.1 Year-wise development of Indore city

S. no.	Year	Area (in sq. km.)	Population
1.	1800	4.0	20,000
2.	1850	9.0	45,000
3.	1900	13.4	99,880
4.	1950	17.2	310,859
5.	2000	131.1	1,600,000
6.	2011	526	2,167,447

Source: Town and Country Planning, Indore

national growth rate, which is 25.7 %. Female population is 912 per 1000 males. The process of growth and development of the city has been as a religious route, as a fertile agricultural land, as an industrial town, and as a business city of Madhya Pradesh.

Indore is the largest city of Madhya Pradesh state in India. The fast development and growth of industries within Indore and its neighboring areas is expected to continue attracting migration for better opportunities (Mishra and Pandit 2013).

11.5.4 Indore Crime Status

There is high pressure of price hike, maintaining socioeconomic status, and moreover survival in a growing megacity like Indore. This can be a major reason among others related to society, culture, family systems and political influences, etc. which contribute to constantly rising crime rates in Indore city. As per the latest published report by National Crime Record Bureau, Indore city's crime rate is second highest in India, the rate being 809.9 % which is second after Coimbatore with crime rate 834.3 %. More precisely, out of the total crimes registered in 53 cities of India, 3.2 % crimes took place in Indore city (Dainik Bhaskar Indore, 2nd August 2014).

11.5.5 The Study Area

For analyzing the role of environmental design and urban regeneration on social sustainability, four locations of the city have been selected. The four locations form two sets on the basis of social status of residents (income level and education). The locations are:

- Janjeerwala Square
- Sapna Sangeeta
- Bhagirath Pura
- Nehru Nagar

11.5.6 Analysis

11.5.6.1 New Palasia

- It is a colony of mixed use, i.e., residential and commercial.
- Traffic circulation is high.
- Predominantly mid-rise buildings in gated colonies with around 2 m high boundary walls.

Fig. 11.1 Janjeerwala

Fig. 11.2 Sapna Sangeeta

- Main road is 20 m wide and internal roads of 10 m width.
- Eateries contribute to high footfall throughout the day and till late night (Fig. 11.1).

11.5.6.2 Sapna Sangeeta

- Buildings facing main road have shops in the ground floor with residences and offices above.
- Main road has high traffic circulation.
- Width of main artery is 15 m.
- Boundary walls of individual houses are 1.5 m high.
- Off-street parking of 3 m is provided (Fig. 11.2).

Fig. 11.3 Bhagirath Pura

Fig. 11.4 Nehru Nagar

11.5.6.3 Bhagirath Pura

- It is a colony which came up for settling mill workers of lower social status.
- The road layout is unorganized and buildings are of varied sizes.
- People have adopted their own security means.
- Roads are unsafe with antisocial activities and hence the area is crime prone.
- Waste management is very poor (Fig. 11.3).

11.5.6.4 Nehru Nagar

- It is a colony designed for mill workers' residential use.
- Most people are educated and for livelihood they have made shops in ground floors.
- Main road is 12 m wide and internal roads are of 5 m.
- Crime rate has come down as the shopkeepers have full-time surveillance over the colony (Fig. 11.4).

11.6 Research Methodology

11.6.1 Objectives

(i) To study relationship of urban regeneration and social sustainability
(ii) To study the impact of environmental design on social sustainability

Quantitative research design is developed for the two objectives. This is an effort of exploration of Indore city on identified four locations – Bhagirath Pura, Nehru Nagar, Sapna Sangeeta, and Janjeerwala.

11.6.2 Unit of Analysis

For conducting exploratory study and finding out urban regeneration of individual location, citizens' viewpoints are very important, as found from previous research and researcher's experience. Hence individual citizen is taken as a unit of analysis. Data is collected from occupants of settlement who are staying there since last two decades. Twenty years can be considered as a module of human life. Also a person of 15 years and above has developed judgmental sense; this is the reason behind keeping the respondent's minimum age as total of 15 and 20, which is 35.

11.6.3 Source of Information

Primary data is collected from the citizens with the help of a questionnaire.

11.6.4 Sampling Design

(i) *Population*: All the citizens living in the identified four locations of Indore city are the universe of this study (Fig. 11.5).
(ii) *Sampling frame*: All individual citizens of four locations of Indore city. The citizens who are living in Indore since last 20 years.
(iii) *Representative sampling unit*: The study is carried out only at four locations of Indore city, which are Bhagirath Pura, Nehru Nagar, Sapna Sangeeta, and Janjeerwala. Representative sampling unit is an individual.
(iv) *Sampling method*: Non-probability convenient sampling method is used.
(v) *Sampling size*: The questionnaire was administered to 320 respondents, 80 citizens from each of the four locations of Indore city. Considering the whole of Indore city, the universe size is too large as the city can be demographically

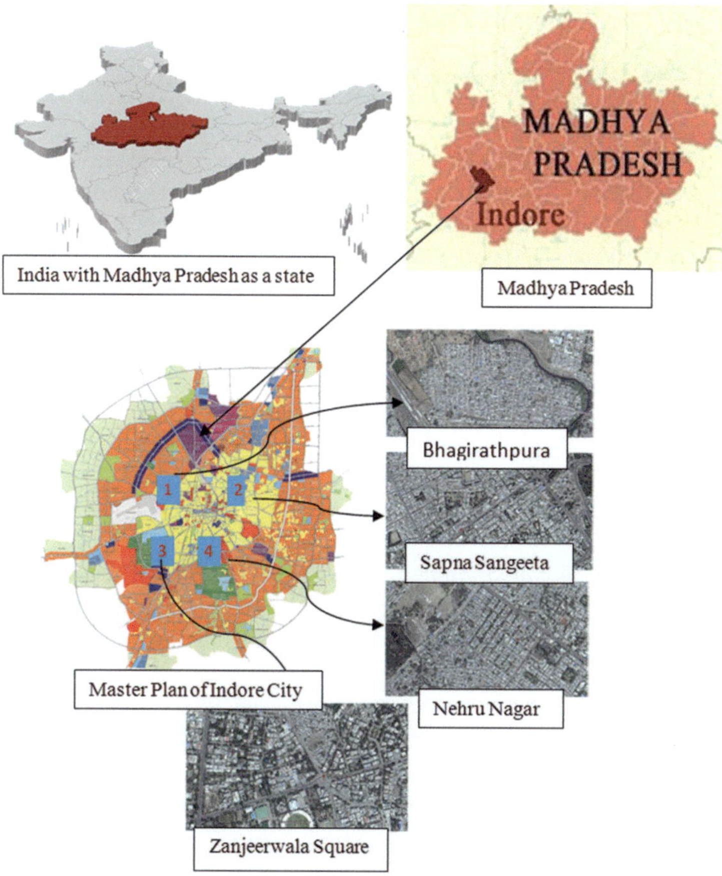

Fig. 11.5 Location of Indore city in India and Google maps of four areas under analysis

divided into many segments. The study is restricted to Bhagirath Pura, Nehru Nagar, Sapna Sangeeta, and Janjeerwala. The respondents are selected from the age group of 35 years and above.

(vi) *Sampling media*: An exhaustive survey by method of personal interview was conducted for collecting data. This was carried out by a student team comprising of 16 students from B. Arch. course of School of Architecture, IPS Academy, Indore.

(vii) *Questionnaire design*: A questionnaire has been designed for the study. In the first phase of design, a list of parameters has been prepared from previous research. For each parameter, variables have been identified from literature review and researcher's experience. Each variable have been transformed into a research question. This list is then converted into a questionnaire for measuring the response of respondents on a scale of 1–5.

11.6.5 Measurement of Variables

The study finds the consent level in the relevance and importance given to the regenerations in various parameters of development. Likert scale is being used to measure the responses of the respondents. (A **Likert scale** is a psychometric scale commonly involved in research that employs questionnaires. It is the most widely used approach to scaling responses in survey research. The scale is named after its inventor, Mr. Rensis Likert). This survey was conducted on general citizens, irrespective of their position in the neighborhood. Thus it was not expected that they will find too minute difference on the levels of each question. This is the reason that scale of 5 levels is used. This scale identifies consent level on each question from 1 to 5 as strongly disagree to strongly agree.

11.6.6 Data Analysis Tools

(i) *Study of relationship of urban regeneration and social sustainability*: Structure equation modeling is used to develop model. Amos software of SPSS 21 is used for data analysis. Relationship is found through simple regression analysis of linear type.
(ii) *Study of impact of environmental design on social sustainability, comparative study of four locations of Indore city*: Structure equation modeling is used to develop model. Amos software of SPSS 21 is used for data analysis. Relationship is found through simple regression analysis of linear type.

11.6.7 Hypotheses

Hypotheses for two objectives are:

Ho 1: There is no significant relationship between urban regeneration and social sustainability.
Ho 2: There is no significant relationship between environmental design and social sustainability.

The first hypothesis (Ho 1) is tested by regression analysis. The assumption here is that social sustainability is dependent on urban regeneration. Simple regression analysis of linear relationship was carried out. The SPSS 21 outcome by using enter method of relationship analysis suggest that R^2 value is 0.819. This value is good to accept the model and preposition that social sustainability is dependent on urban regeneration. Thus further analysis of relationship is to be carried out. The R^2 value represents the variance explained of the model. The maximum value of R^2 can be 1 and value more than 0.7 represents acceptable model (Fig. 11.6).

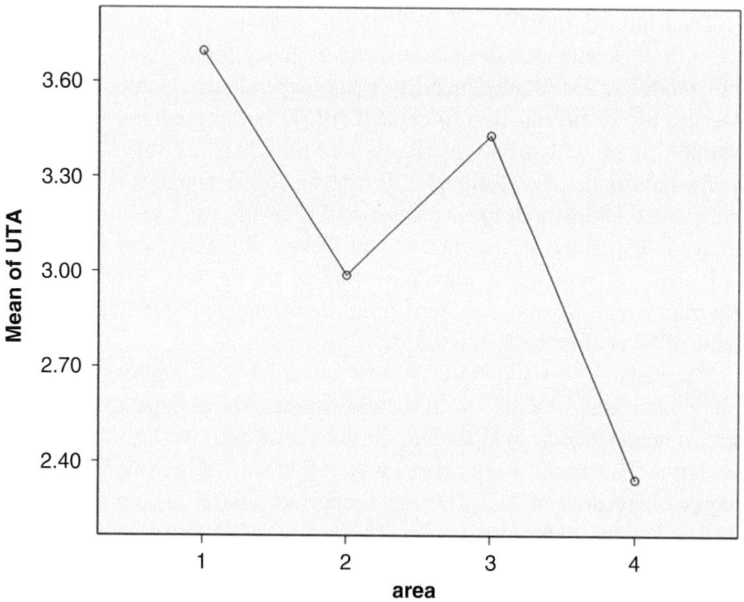

Fig. 11.6 ANOVA analysis of urban regeneration

The Fisher's test and ANOVA analysis at 95 % of confidence limit shows that F value is 1436.095 at total degree of freedom of 318. The significance value reported is 0.000. The magnitude of significance value is less than 0.05. This reflects that linear relationship exists between the variables. The coefficient of the regression as calculated by SPSS reflects that constant value is 1.060 and coefficient of urban regeneration is reported as 1.231:

$$SS(\text{Social Sustainability}) = 1.060 + 1.231UR(\text{Urban Regeneration})$$

The relationship as expressed by these values is very significant as small change in urban regeneration will bring out larger change in social sustainability.

Thus, we can conclude that the social sustainability is dependent on urban regeneration.

The second hypothesis (Ho 2) is tested by regression analysis. The assumption here is that social sustainability is dependent on environmental design. Simple regression analysis of linear type was carried out. The SPSS 21 outcome by using enter method of relationship analysis suggest that

(i) For model 1, when social sustainability is the dependent variable and urban regeneration is the independent variable, R^2 value is 0.819. This means about 81.9 % of the variation is social sustainability which is explained by urban regeneration. The regression equation appears to be very useful since the value of R^2 is close to 1.

Then, mixed variables of environmental design are introduced to see the effect of environmental design on social sustainability.

(ii) For model 2, social sustainability is the dependent variable; Environmental Design for Territorial Reinforcement (EDT) is the predictor or independent variable along with urban regeneration as prior existing variable.

As Environmental Design for Territorial Reinforcement is introduced, R^2 change is 0.855, which means about 85.5 % of the variation in social sustainability is explained by Environmental Design for Territorial Reinforcement, and also the regression equation appears to be very useful for making Environmental Design for Territorial Reinforcement predictions since the value of R^2 is close to 1, i.e., 0.855.

The Fisher's test and ANOVA analysis at 95 % of confidence limit shows that F value was 1436.095 with social sustainability as dependent variable and urban regeneration as independent variable; after introduction of Environmental Design for Territorial Reinforcement, the F value has become 933.500 at total degree of freedom of 319. The value reported is 0.000. The magnitude of significance value is less than 0.05. This reflects that linear relationship exists between the variables. The coefficient of the regression as calculated by SPSS reflects that constant value is 1.269 and coefficient of urban regeneration has reduced from 1.231 to 0.830 due to introduction of Environmental Design for Territorial Reinforcement, meaning Environmental Design for Territorial Reinforcement has a lot of impact on social sustainability, from the regression equation:

$$\text{Social Sustainability} = 1.269 + 0.830\,\text{UR}\left(\text{Urban Regeneration}\right) + 0.199\text{EDT}$$
$$\left(\text{Environmental Design for Territorial Reinforcement}\right)$$

(iii) For model 3, social sustainability being the dependent variable and Environmental Design for Natural Surveillance (EDN) as independent variable are introduced along with preexisting independent variables – urban regeneration and Environmental Design for Territorial Reinforcement.

As Environmental Design for Natural Surveillance is introduced, R^2 change is 0.886 which means about 88.6 % of the variation in social sustainability is explained by Environmental Design for Natural Surveillance; also the regression equation appears to be very useful for making predictions as the value of R^2 is close to 1, i.e., 0.886.

The Fisher's test and ANOVA analysis at 95 % of confidence limit shows that with urban regeneration and Environmental Design for Territorial Reinforcement as predictors, F value was 933.50; after introduction of Environmental Design for Natural Surveillance as one more predictor (independent variable) along with urban regeneration and Environmental Design for Territorial Reinforcement, the F value is 818.098 at total degree of freedom 319. The significance value reported is 0.000. The magnitude of significance value is less than 0.05; this reflects that linear relationship exists between the variables. The coefficient of the regression as calculated by SPSS reflects that

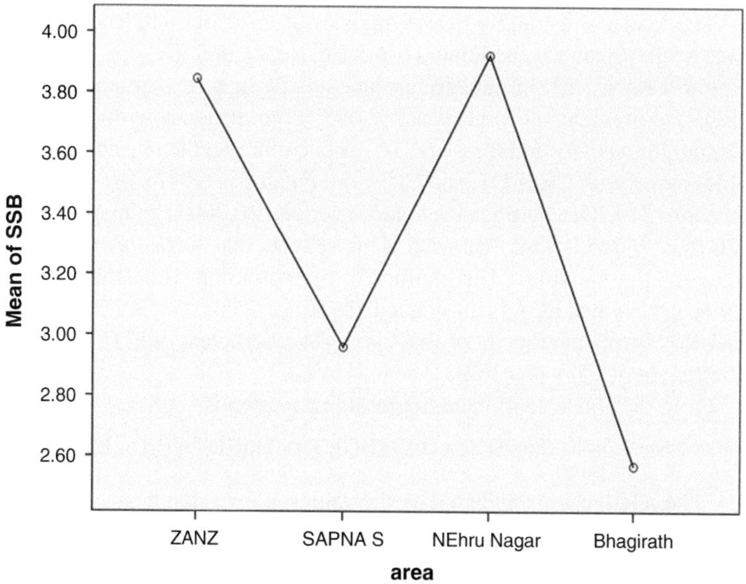

Fig. 11.7 ANOVA analysis of social sustainability

when Environmental Design for Natural Surveillance is introduced, the value of constant is 1.291, coefficient of urban regeneration is 0.636, and coefficient of Environmental Design for Natural Surveillance is 0.270 (Fig. 11.7).

From this coefficient, regression equation is

$$\begin{aligned}\text{SS}(\text{Social Sustainability}) &= 1.291 + 0.636\,\text{UR}(\text{Urban Regeneration}) \\ &+0.270\text{EDN}(\text{Environmental Design for Natural Surveillance}) \\ &-0.015\text{EDT}(\text{Environmental Design for Territorial Reinforcement}).\end{aligned}$$

The relationship as expressed by the values is very significant as small change in Environmental Design for Natural Surveillance will bring out larger change in social sustainability. Thus, it proves that Environmental Design for Natural Surveillance has a significant effect on social sustainability.

(iv) For model 4, social sustainability as dependent variable and Environmental Design for Activity and Image (EDA) as independent variable are introduced along with preexisting independent variables – urban regeneration, Environmental Design for Territorial Reinforcement, and Environmental Design for Natural Surveillance.

As Environmental Design for Activity and Image is introduced, R^2 change is 0.897 which means about 89.7 % of the variation in social sustainability is explained by Environmental Design for Activity and Image, and also the regression equation appears to be very useful for making predictions, as the value of R2 is close to 1, i.e., 0.897.

The Fisher's test and ANOVA analysis at 95 % of confidence limit shows that with urban regeneration (UR), Environmental Design for Territorial Reinforcement (EDT), and Environmental Design for Natural Surveillance (EDN) as predictors, F value was 818.098; after introduction of Environmental Design for Activity and Image (EDA) as one more predictor (independent variable) along with UR, EDT, and EDN, the F value is 688-084 at total degree of freedom 319. The significance value reported is 0.000. The magnitude of significance value is less than 0.05. This reflects that linear relationship exists between the variables. The coefficient of regression as calculated by SPSS reflects that when EDA is introduced, the value of constant is 1.384, coefficient of UR is 0.465, coefficient of EDN is 0.194, coefficient of EDT is –0.036, and coefficient of EDA is 0.162.

From the above coefficient, regression equation is

$$SS = 1.384 + 0.465UR + 0.194EDN - 0.036EDT + 0.162EDA$$

The relationship expressed by the values is very significant, as any change in EDA will bring about larger change in SS. Thus it is clear that EDA has significant effect on SS.

(v) For model 5, social sustainability as dependent variable and Environmental Design for Milieu and Accessibility (EMA) as independent variable are introduced along with preexisting independent variables – urban regeneration (UR), Environmental Design for Territorial Reinforcement (EDT), Environmental Design for Natural Surveillance (EDN), and Environmental Design for Activity and Image (EDA).

As Environmental Design for Milieu and Accessibility (EMA) is introduced, R^2 change is 0.906 which means about 90.6 % of the variation in social sustainability is explained by Environmental Design for Milieu and Accessibility (EMA), and also the regression equation appears to be very useful for making predictions, as the value of R2 is close to 1, i.e., 0.906.

The Fisher's test and ANOVA analysis at 95 % of confidence limit shows that with UR, EDT, EDN, and EDA as predictors, F value was 688.084; after introduction of EMA as one more predictor (independent variable) along with UR, EDT, EDN, and EDA, the F value is 604.116 at total degree of freedom 319. The significance value reported is 0.000. The magnitude of significance value is less than 0.05. This reflects that linear relationship exists between the variables. The coefficient of regression as calculated by SPSS reflects that when EDA is introduced, the value of constant is 1.370, coefficient of UR is 0.512, coefficient of EDN is 0.204, coefficient of EDT is 0.007, coefficient of EDA is 0.169, and coefficient of EMA is –0.096.

From the above coefficient, regression equation is

$$SS = 1.370 + 0.512UR + 0.204EDN + 0.007EDT + 0.169EDA - 0.096EMA$$

The relationship expressed by the values is very significant, as any change in EMA will bring about larger change in SS. Thus it is clear that EMA has significant effect on SS (Fig. 11.8).

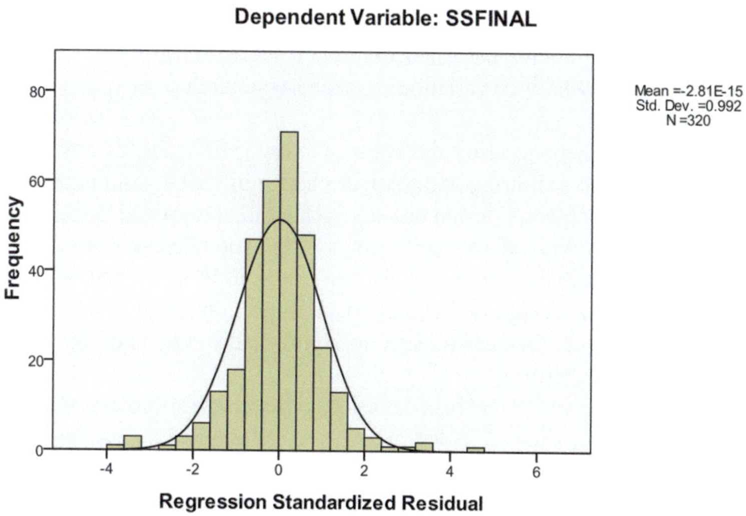

Fig. 11.8 Regression histogram

It is observed that when the variables of environmental design were introduced, the F value got reduced; the coefficient also got changed. Hence, it can be concluded that the environmental design (ED) takes up the impact on social sustainability.

Hereby, the null hypothesis is rejected as significance value is 0.000 and model does improve the prediction.

11.7 Result

Data evolved is analyzed and the results prove that the regenerated urban places with consciously designed environment have low crime rates, hence are socially more sustainable.

According to the final model,

$$SS = 1.370 + 0.512\,UR + 0.204\,EDN + 0.007\,EDT + 0.169\,EDA - 0.096\,EMA$$

11.8 Conclusion

Researcher's preposition is that urban regeneration and environmental design act together. According to the final model,

$$SS = 1.370 + 0.512\,UR + 0.204\,EDN + 0.007\,EDT + 0.169\,EDA - 0.096\,EMA$$

All constructs of environmental design have their impact on social sustainability; similarly, urban regeneration has impact on social sustainability.

As per R^2 values, 90.6 % of variation in social sustainability is explained by environmental design.

The relationship expressed by the values is very significant, as any change in environmental design will bring about larger change in social sustainability. Thus, it is clear that environmental design has significant effect on social sustainability.

It is observed that when the variables of environmental design were introduced, the F value got reduced; the coefficient also got changed. Hence, it can be concluded that the environmental design (ED) takes up the impact on social sustainability.

Hereby the null hypothesis is rejected as significance value is 0.000 and model does improve the prediction.

The areas under research had been so selected that they form two sets with similar social status. Study showed it clearly that New Palasia and Sapna Sangeeta with similar social conditions differ in crime rates due to difference in the types of environmental planning being adopted. Same result has evolved in the set of Bhagirath Pura and Nehru Nagar, both having similar social structure. Hence, it is clear that environmental design plays an important role in promotion or restriction of crime in a certain location. Also urban regeneration and environmental design both are directly proportional to social sustainability of any urban settlement.

References

Cozens PM, Saville G, Hillier D (2005) Crime prevention through environmental design (CPTED): a review and modern bibliography. Prop Manag 23(5):328–356

Davis (1995) Fortress Los Angeles: the militarization of urban spaces, metropolis. McMillan, Los Angeles

Glasson J, Cozens P (2010) Making communities safer from crime: an undervalued element in impact assessment. Environ Impact Assess Rev

Graham S (2005) The city. Columbia, Oxford

John Malcom MG (1985) Central India. Government of India, Bhopal

Lefebre H (1996) Right to the city. Translated by Kofman and Lebas, pp. 131–142

Mishra SA, Pandit R (2013a) Comparative analysis of Bhopal, Indore and Gwalior's urban transformation. Int J Combined Res Dev 1(5):1–9

Mishra SA, Pandit RK (2013b) Urban transformation and role of architecture towards social sustainability. Int J Eng Res Dev 5:16–20

Newman O (1972) Defensible space; crime prevention through urban design. McMillan, New York

Rossi A (1984) The architecture of the city. MIT Press, Cambridge, MA

Wright FL (1953) "Some Aspects of the Past and Present of Architecture" the future of architecture. Horizon Press, New York

Chapter 12
Vernacular Architecture: A Sustainable Approach

Vijayalaxmi K. Biradar and Shashikala Mama

Abstract The fundamental purpose of buildings is to provide a comfortable living environment protected from the extremes of climate. In this regard vernacular buildings evolved gradually to meet environmental, socioeconomic, and sociocultural characters of the society to meet the changed lifestyle over a period of time. These buildings obtained deep harmonization with site surroundings and had a minimal environmental impact. The concept of sustainability, which arose during the 1980s, evolved very rapidly from an ecologically friendly approach to a series of rather high-tech and expensive responses. Nevertheless, vernacular heritage throughout the world was, and is, very much alive and still plays an active role in present society and its architecture.

The main objective of this paper is to understand the principles and strategies for sustainability from vernacular heritage and integrate those in present society. The methodology adopted is documentation of the vernacular built heritage of Kalaburagi, a hot and dry climate region of Karnataka State, and analyzed based on environmental, sociocultural, socioeconomic, planning, and energy conservation aspects. This paper concludes with an appreciation of principles of vernacular architecture to integrate them in the present scenario.

Keywords Vernacular architecture • Climate • Planning • Environmental • Sustainable development • Courtyard

12.1 Introduction

Vernacular architecture is composed of traditional buildings, which represent gradually evolved response to meet environmental and climatic constraints, as well as the socioeconomic and cultural characters of societies to meet changed lifestyle over a period of time (Broadbent 1975). Additionally, the materials and

---

V.K. Biradar (✉) • S. Mama
Poojya Doddappa Appa College of Engineering,
Kalaburagi (Gulbarga) 585 102, Karnataka, India
e-mail: viju1009patil@yahoo.co.in; mamashashikala@gmail.com

F. Seta et al. (eds.), *Understanding Built Environment*, Springer Transactions in Civil and Environmental Engineering, DOI 10.1007/978-981-10-2138-1_12

architectural components used are climate responsive and tailored according to distinct locations, and have therefore adapted to geographic and topographical features, as well as to local climate. This type of architecture normally presents a good climate adaptation and provides good thermal comfort due to the choice of natural materials and bioclimatic features adapted to the environment. Besides, it is a cost-effective architecture, both in economic and social terms, self-sufficient as regards to natural resources, and with a low environmental impact and, therefore, with a sustainable input.

The concept of sustainability of our project is to provide an environment-friendly building with respect to improving the quality of social life of the people, enhancing the economy, and reducing the impact of the buildings on the environment.

To achieve sustainability the major parameters considered are:

- Environmental sustainability
- Sociocultural sustainability
- Socioeconomic sustainability
- Planning concepts
- Energy conservation

12.2 Environmental Sustainability

KALABURAGI (GULBARGA) means "stony land" in Kannada language. Kalaburagi District is situated in the northern part of Karnataka State and is about 353 miles southeast of Mumbai (Fig. 12.1). It lies between longitude 76° 04′ and

Fig. 12.1 Location of Kalaburagi (Gulbarga)

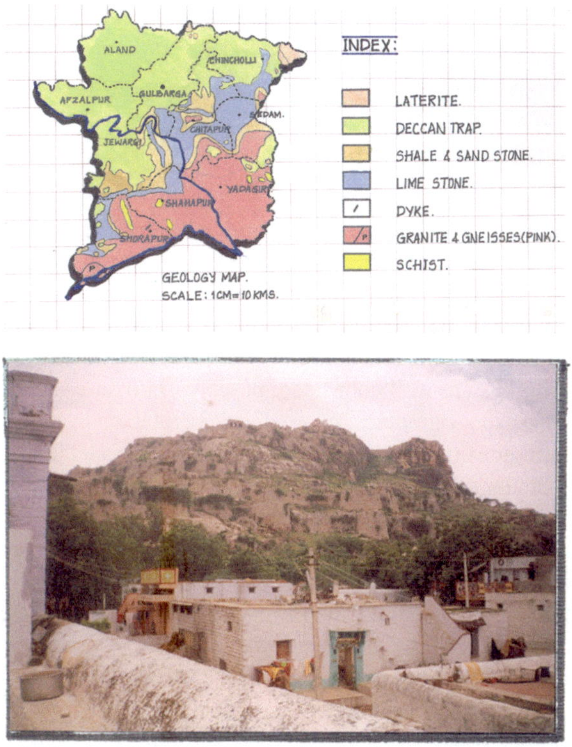

Fig. 12.2 Geology of Kalaburagi (Gulbarga)

77° 42′ and latitude 16° 12′ and 17° 46′. The total area is 6,271.2 miles^2. The main rivers are Krishna and Bhima (Bindoo 1942).

Local climate, topography, and natural resources of this region are the starting point from which the architectural organism conceived (Oliver 1997).

The entire district is situated in Deccan Plateau. A range of hills from North to SE provides hard and compact stone formations. The remaining part of this district is flat.

The district consists of varieties of stones like Deccan Trap, sedimentary rock, and Archean rock (Fig. 12.2). These are natural materials, locally available, often organic, renewable and biodegradable. The longevity of stone houses of Kalaburagi is mainly because they are flexible, durable, and resilient to climate change. The "cradle to cradle" approach refers to a building that can be reconstructed by using recycled materials.

The climate of this district is hot and dry. The temperature ranges from 5 to 45 °C; winds are generally light to moderate and rainfall about 776.5 mm annually (Koenisberger 2001). This bright sunlight, excessive heat, scanty rainfall, and abundance of stone have affected a great deal of domestic architecture.

Vernacular communities of Kalaburagi have assured the livability and comfort of their homes by taking advantage of the topography, climate, and natural resources.

12.3 Sociocultural Sustainability

The great importance of sociocultural heritage of Kalaburagi region is reflected in proper linkages between the cultures, nature, space, identity of the local society, and ability for creating the best living conditions.

The homes of ancestors in Kalaburagi region were so harmoniously integrated in the landscape as a geographic and cultural space and as the mirror of man's history and life. The buildings are well connected; peaceful, human-scale architecture that fits into sites respects topographies, mineral, and vegetal environments.

Compared with housing that appears to be fragile, degradable, and built for one generation, vernacular architecture of this area proclaims the reassuring evidence of its sturdiness (miniature fort) and of its environmental sustainability.

In vernacular built heritage, a tulasi or jasmine plant and water are considered as elements of cultural values.

The vernacular human settlements in this region reflect inhabitant's desire to maintain the conditions of social organization so as to live together peacefully. The social and cultural heritage reveals a high level of skill in terms of adapting local resources and using them wisely (Corria et al. 2014).

The social and cultural elements of vernacular architecture (Fig. 12.3) are also reflected in the building language.

12.4 Socioeconomic Sustainability

Vernacular architecture of rural and isolated areas of Kalaburagi reflects a balance between energy saving, tradition, the environment, and the social parameter.

The development of effective strategies for sustainable development is achieved in this region by supporting local communities to be more self-sufficient, by sustaining local production, by optimizing local materials, and by choosing to work with collective efforts (Corria et al. 2014).

In sustainable economic systems, farm houses in villages of this region are adopted an efficient management of resources, including people, land, animals, equipment, buildings, materials, resources, etc. to avoid waste and save energy (Corria et al. 2014).

The advantage of exchange, trade, and even barter system encourages to produce, to sell, and to purchase in villages of this region to achieve sustainable economies capable of being more self-sufficient – as mentioned by Oliver (1997).

Socioeconomic balance is achieved in this region through minimum effort, reduced movement, efficient work, effective management of the available resources, and collaborative efforts to ensure maintenance over time (Fig. 12.4) because of local resources (Corria et al. 2014).

Fig. 12.3 Cultural features of the area (stone balcony, main door, roof, front facade)

Fig. 12.4 Showing socioeconomic balance

12.5 Planning Concepts

The courtyard house has been the most prevalent architectural typology in vernacular architecture of hot-dry climate of Kalaburagi region. The spatial organization of these dwellings is defined by a central open-air space, the core of the house on which the rooms depend and receive light and air.

The typological configuration of the house with one or more courtyards has been one of the common features of the housing cultures in this area. House types are based on economy, trades, and social status (Fig. 12.5).

The openings to the outside are scarce, small, and often protected with lattice works, and balconies at higher level allow seeing without being seen.

The central open courtyard is traditionally the space for sharing, socialization, and household labor.

The elements of the courtyard spatial system are spaces linking the house with the road (space filter), spaces used for domestic activities and family meetings (courtyard and terraces), pathways (porticoes and galleries), and spaces for residence (rooms). All these elements are arranged around the central courtyard and correspond to an ordered sequence determined by the use of space and the need (Fig. 12.6).

12.5.1 Materials and Construction Techniques

Proper use of right material in the right place and use of traditional technologies make the built form in vernacular buildings of this region energy efficient and sustainable.

High thermal inertia found in vernacular constructions of Kalaburagi region that use stone and lime mortar in their walls and ceilings (Fig. 12.7). The building system consists of high thermal mass finished with lime coating which has an antibacterial functionality and solar radiation protection.

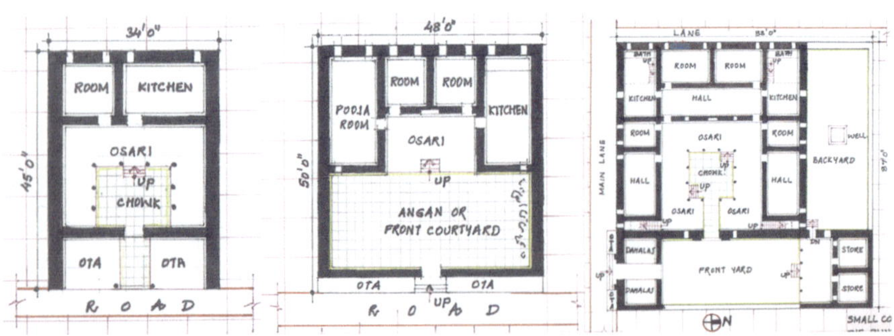

Fig. 12.5 Types of houses based on economy, trade, and social status

Fig. 12.6 Planning features of a vernacular house: cowshed, plan of the house, rooms around courtyard, and main entrance

Fig. 12.7 Shows use of materials and construction techniques used in walls, roof, and floor

12.6 Energy Conservation

Vernacular architecture plays a vital role toward sustainability by developing climate-responsive structures.

12.6.1 Orientation

The buildings are oriented east-west directions and make the spatial planning more perfect to control its environment with maximum comfort in all seasons due to shorter walls facing sun (Fig. 12.8).

12.6.2 Microclimate

Planting trees to shade walls and mutual shading of the building is more beneficial in reducing the heat gain (Fig. 12.9). Deciduous plants found to be more beneficial in this region.

12.6.3 Integrating Indoor and Outdoor Living

The hierarchical positions of courtyard, verandah, and inner spaces provide an organization of activities in relation to climate (Fig. 12.10). The verandah, acting as buffer space, helps in slow penetration of heat from courtyard to indoor spaces (Ali Asif 2012).

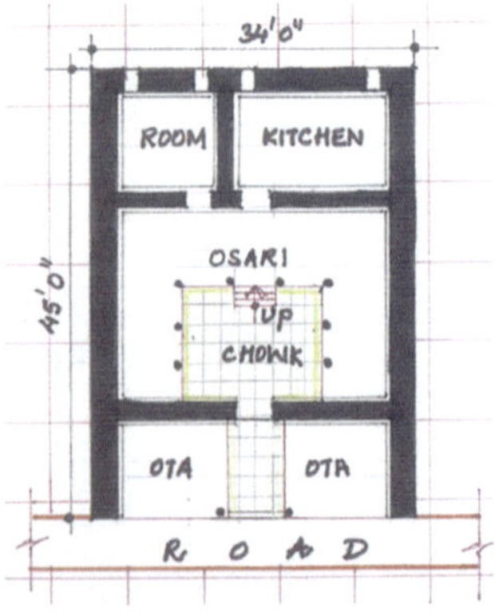

Fig. 12.8 Building orientation

Fig. 12.9 Shading and mutual shading of buildings and street

Fig. 12.10 Relationship between indoor and outdoor spaces

12.6.4 Thermal Mass

The thermal inertia of the walls and roof keeps interiors cool due to larger time lag because of thick walls and roof as well as use of low thermal conductivity materials (Fig. 12.11).

Fig. 12.11 Thermal insulation of wall and roof

Fig. 12.12 Centrally placed courtyard provides a comfortable place for living

12.6.5 Courtyard

In courtyard houses, due to the air convention property, warm air is less dense than cool air which will rise, and cooler air replaces hot air from other openings generating a continuous air current (Philokyprou and Michael 2012) (Fig. 12.12).

During the day inhabitants use different parts of the courtyard according to the movement of the shade.

Trees and plants also play a very important role in helping to induce airflow and provide shade on hotter days in courtyard houses, and water will help cool the air through evaporation and can have a calming and soothing effect.

Fig. 12.13 Shading devices are used to protect from heat penetration

Fig. 12.14 Natural ventilation through openings

12.6.6 Shading Devices

The natural cooling is achieved by providing shading devices such as projecting stone balconies, sun shades, verandah, and brackets (Fig. 12.13).

12.6.7 Natural Ventilation

Natural ventilation is the best result of differential wind forces between outside and inside air which maintains comfort inside (Fig. 12.14).

12.7 Result

From the study it is clear that vernacular architecture of this region establishes an adaptable responsive spatial organization by using adequate principles of design, expertise in the integration to the place, smart ways of producing natural renewable energy, and reduction of cost and transportation.

So, vernacular buildings have become a guide for sustainable development. The findings from this study suggest further research to explore guidelines for designing buildings for sustainable development in the present era and future world.

12.8 Conclusion

Vernacular architecture becomes an important case to analyze for better understanding as well as for recognizing the value of self-sustaining principles.

Vernacular heritage represents a great resource that has significant potential to define principles for sustainable design. The fact is that vernacular architecture and its methods and strategies are undervalued and seldom applied in recent building trends (Wahid 2012).

The challenge is to find out how this achieved knowledge, skills, and experience of the world's vernacular builders may be fruitfully applied in a modern context.

In achieving sustainability, the priority must be given to minimize the depletion of natural resources, safeguarding ecosystems, identifying and developing new materials and construction techniques to minimize waste, and nonpolluting for the benefit of our future generations.

The role of an architect is essential to the building design: the most environment-friendly and energy-efficient building must also be functional, durable, and aesthetically pleasing.

"The principles that produced the vernacular solutions must be respected. This is the only way modern architecture can surpass, in human and ecological quality, the achievement of vernacular architecture in all the regions of the world" (Fathy 1986).

References

Ali Asif (2012) Passive cooling and vernacularism in Mughal buildings in North India: a source of inspiration for sustainable development – article history, Architecture Section (University Polytechnic), Aligarh Muslim University, INDIA.

Bindoo DD (1942) A critical study of the old domestic architectural buildings of Gulbarga, (A thesis report). School of Architecture Bombay

Broadbent G (1975) Design in architecture, architecture & the human sciences. Wiley, London

Corria M, Dipasquale L, Mecca S (2014) VERSUS heritage tomorrow, vernacular knowledge for sustainable architecture. Firenze University Press, Firenze

Fathy H (1986) Natural energy and vernacular architecture: principles and examples with reference to hot arid climate. The University of Chicago Press, Chicago
Koenisberger TG (2001) Manual of tropical housing and building. Orient Longman Ltd, Chennai
Oliver P (1997) Encyclopedia of vernacular architecture of the world, vol 1. Cambridge University Press, Cambridge
Philokyprou M, Michael A (2012) Evaluation of the environmental features of vernacular architecture-a case study in Cyprus. University of Cyprus, Department of Architecture, Nicosia
Wahid A (2012) Adaptive vernacular options for sustainable architecture. J Int Soc Study Vernac Settl 2:74–85

Chapter 13
Urban Regeneration and Sustainability: Importance of Sustainable Transport Systems in the Concept of Eco-City

Minu Joshi

Abstract Transportation and mobility are essential prerequisites for a sustainable development like eco-city. Statistics show that billions of dollars are invested in developing and maintaining transport systems all over the globe. The aim of this paper is to explore the importance of sustainable transport in the concept of eco-city using appropriate precedents and discuss its benefits. Case studies show that a well-designed and efficient transport system contributes to the economic growth and helps in refining the quality of life of the people, thereby becoming one of the most important sectors of urban development. Beyond the economic growth and productivity, transportation is also a major component of urban environmental management as seen in the case of Bad Ischl and Tianjin. Further the paper also looks at the other benefits of sustainable transport like health benefits and societal benefits using examples. The paper concludes by suggesting that sustainable transport plays a key role in supporting the concept of an 'eco-city'.

Keywords Sustainable transport • Transport for eco-city • Environmental impact of transportation • Eco-city

13.1 Introduction

Among the four major sectors of urban development in the concept of eco-city, namely, the 'urban structure, transport, energy, material flow and socio-economy' – transportation takes the most important position due to its influence on all the other sectors. Transport enables mobility of people and goods from place to place. People travel for a variety of reasons every day, for example, to work, for leisure, to

M. Joshi (✉)
SCET, Surat, India

Theory and Practice of Sustainable Design, Cardiff University, Cardiff, UK
e-mail: ecour.ar@gmail.com

F. Seta et al. (eds.), *Understanding Built Environment*, Springer Transactions in Civil and Environmental Engineering, DOI 10.1007/978-981-10-2138-1_13

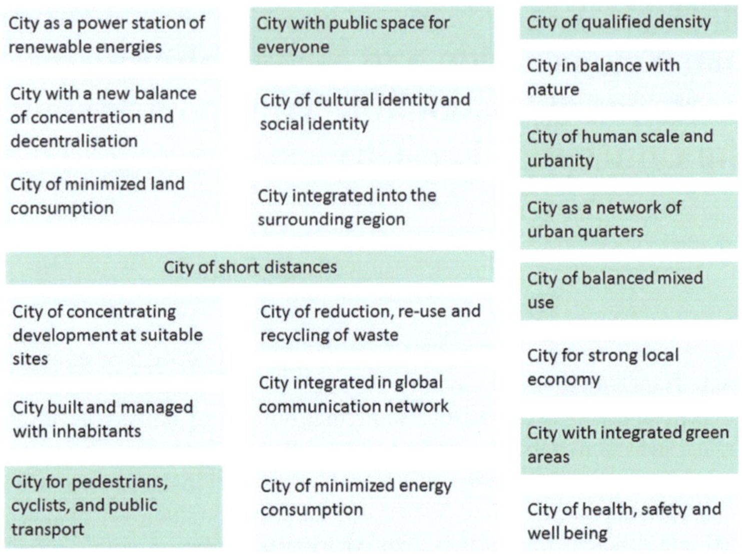

Fig. 13.1 Vision of eco-city – the 'city of' diagram – outcome of the eco-city workshop II (Vienna June 2002) (Source – The Conceptual Framework of the EU Project Eco-city. (n.d.). Retrieved Jan 20, 2016, from http://www.ecocity.szm.com/framework.html)

socialize, for shopping, etc. Apart from being a crucial and an unavoidable facility in the current scenario, transportation has several adverse effects on the environment. Statistics show that transportation alone is responsible for a seventh of greenhouse gas emissions which will eventually have a catastrophic impact on the climate. Increase in use of motorized vehicles in past few decades have resulted into issues like congestion, waste of energy, safety issues, deteriorating quality of life and so on. In spite of these evident issues associated with transport, one cannot negate that transportation is an essential prerequisite for a sustainable development.

A sustainable development like 'eco-city' needs a transportation system that limits emissions and waste generation within the planet's ability to absorb (Black 2010), allows basic access to people, promotes walkability, is cost-effective and supports a vibrant economy (Skala et al.; Spotts 2013). Among the various modes of transportation, walking and cycling (active travel) are the most sustainable means as they are the least energy consuming and have health benefits. Keeping this idea in consideration, urban patterns appropriate for active travel become a key element for designing an eco-city (Coplak et al.). Among the various parameters enlisted under the vision of an eco-city, many of them draw attention towards a sustainable transportation system to improve the liveability (Ecocity 2001). Principles like 'city of accessibility for everyone', 'city of pedestrians and cyclists', city of public transport' and 'city of short distances' all suggest the importance of transportation and connectivity (ref Fig. 13.1). All these points mentioned above suggest a need of high accessibility to the most essential destinations (Skala et al.; Coplak et al.) like the grocery store, school or work place. This also suggests that there is a certain level of population density consideration and mixed land use in the urban planning. Barton et al. (2010) set criteria for accessibility to promote active travel as the most preferred modes of transportation in a home place, neighbourhood, urban district and city. According to this criteria chart, the amenities like schools, health centres and local shops must be located within the radius of 600 m, while amenities like pub, post office, community centre and leisure centre can be located in the radius of 800–1,900 m. Similar proposal is put forth in EU eco-city project, where it is suggested that amenities like playground, pin boxes for delivery of goods and semipublic spaces should be accessible within 2 min of walking distance, while facilities like a bus stop, emergency centres, service points, shops, larger public parks, youth centre and schools should be located within 300–500 m (Skala et al.; Barton et al. 2010). These amenities can be accessible either by active travel or by public transportation modes. Studies show that improving the accessibility and creating pedestrian friendly conditions for daily mobility results into car-free living. There are certain other aspects that favour active travel apart from accessibility as suggested by the sustainable cities institute (transit-oriented development); they are:

1. Urban density
2. Mixed-use spaces
3. Attractive and liveable public spaces, recreational spaces, public parks and integrated green spaces
4. Safety and comfort

13.2 Precedents

There are various models implemented on the basic principles of an eco-city. This paper focuses on three examples that are particularly based on the principles related to transportation suggested in the vision of eco-city:

(A) Vauban, Freiburg (Germany)
    Principle – *Car-free settlement*
(B) Bad Ischl, Austria
    Principle – *City of short distance*
(C) Sino-Singapore Tianjin Eco-city
    Principle – *City of green transport*

13.2.1 Vauban, Freiburg (Germany) – Car-Free Settlement

Aiming to create a neighbourhood that provided a car-free and pedestrian- and bicycle-friendly environment and affordable public transportation, Vauban developed a new mobility concept and street typology (Broaddus 2010). As per this concept, the following elements were considered:

Fig. 13.2 Vauban master plan (Source– Vauban°de (n.d.). Retrieved Jan 20, 2016, from http://www.vauban.de/karte/index.html). *Yellow* in Fig. 13.3 shows the car-accessible road; internal streets are majorly pedestrian

(a) Unbundled parking
(b) Making driving a least convenient option to travel
(c) Making bicycling and walking the most convenient option to travel
(d) Improving accessibility
(e) Concept of 'play street'

As a result of these design principles, Vauban is able to provide children's safety and a car-free environment. This helps further to calm the traffic, reduce the pollution and improve the overall quality of life. Along with restricted entry and movement of private vehicles, providing an effective public transport in the form of tram system caters to the needs of people to travel. By giving priority to pedestrians and cyclists, Vauban has managed to reduce car – ownership (Figs. 13.2 and 13.3).

13.2.2 Bad Ischl, Austria: City of Short Distance

Bad Ischl was designed to be a city of short distances, where the basic amenities are planned within walkable distances, and appropriate urban structures are created for pedestrians. This was mainly possible as the settlement is designed within a radius of 300 m. The key design considerations are as follows:

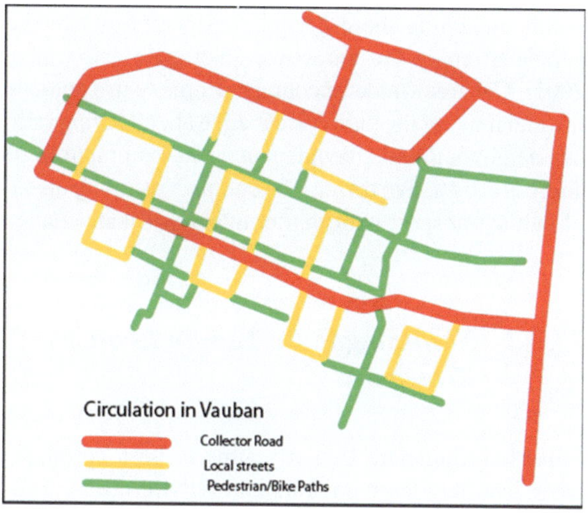

Fig. 13.3 Vauban circulation (Source – Vauban, Freiburg. (n.d.). Retrieved Jan 20, 2016, from https://en.wikipedia.org/wiki/Vauban,_Freiburg). *Red* in Fig. 13.3 shows the car-accessible road; internal streets are majorly pedestrian and bicycle friendly

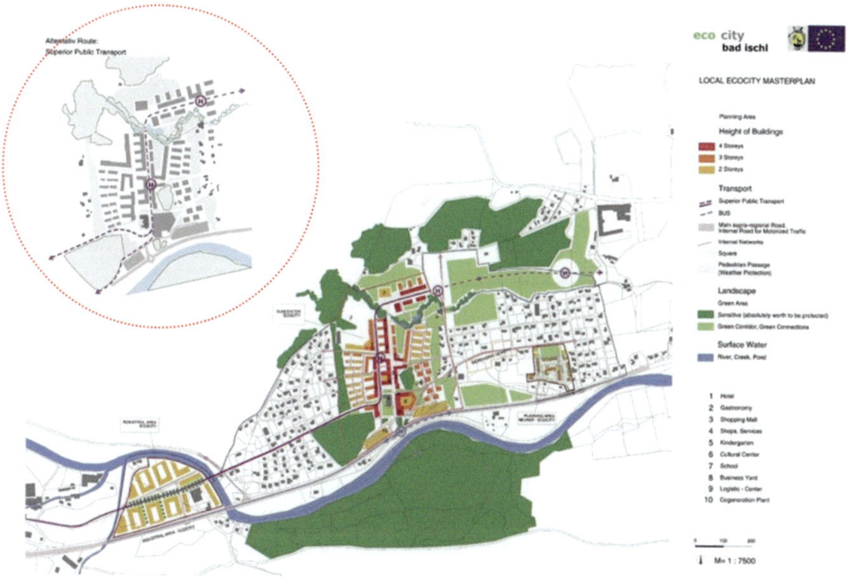

Fig. 13.4 Bad Ischl master plan showing the public transport connectivity (Source: [Brochure]. Retrieved from http://www.upv.es/contenidos/CAMUNISO/info/U0511247.pdf)

(a) Improving accessibility by locating facilities strategically
(b) Designing a balanced mixed use
(c) Designing live able public space
(d) Conserving environment by creating green corridors (Fig. 13.4)

Implementing these design considerations has led to creating more space for people to enjoy the attractive, quiet, safe and wholesome environment (Ecocity 2001). Car-free streets encourage people to use outdoor spaces improving the social life and making the public areas alive and reducing crime rates. As a result of mixed-use neighbourhoods, people can easily commute to jobs, school, shopping, recreation, etc. by active travel. The compact planning also facilitates attractive timetable of public transport with shorter intervals (Skala et al.)

13.2.3 Sino-Singapore Tianjin Eco-City: City of Green Transport

The Sino-Singapore Eco-city aims to be a compact city and provide integrated work-live-play-learn environment (UNEP 2013). This particular eco-city is based on the concept of establishing the harmony between society, economy and environment, the three pillars of holistic sustainability. The aim is to develop a city that supports a vibrant economy and is practical and replicable. Sustainable transport is a key characteristic of Sino-Singapore Eco-city. It aims to do the following:

(a) Increase trips via public transport and active travel.
(b) Avoid conflict between pedestrians, cyclists and vehicles by designing separate network for motorized vehicles.
(c) Priority given to public transport over private vehicles.
(d) Incorporating extensive green (vegetation) and blue (water) networks to provide a better environment for people to walk.

In case of Sino-Singapore Eco-city, the aim was to increase the use of public transport and nonmotorized modes. This is efficiently done by planning a 'cell' that is human centric 400 × 400 m module. These cells are designed to achieve comfortable walking distances, maximize accessibility and provide basic amenities at a comfortable distance. This system helps to streamline the urban development making use of cells, e.g. four cells make one neighbourhood (UNEP 2013). This helps the planners to provide basic amenities and facilities in a tiered manner. Design considerations, like provision of green spaces, generate an appealing environment for people and act as green lung at the central core of the city (Figs. 13.5 and 13.6).

13.3 Overall Benefits of Sustainable Transport

Apart from the benefits discussed for each precedent, there are some other benefits of promoting sustainable transport. These are categorized as health benefits, economic benefits, societal benefits and environmental benefits.

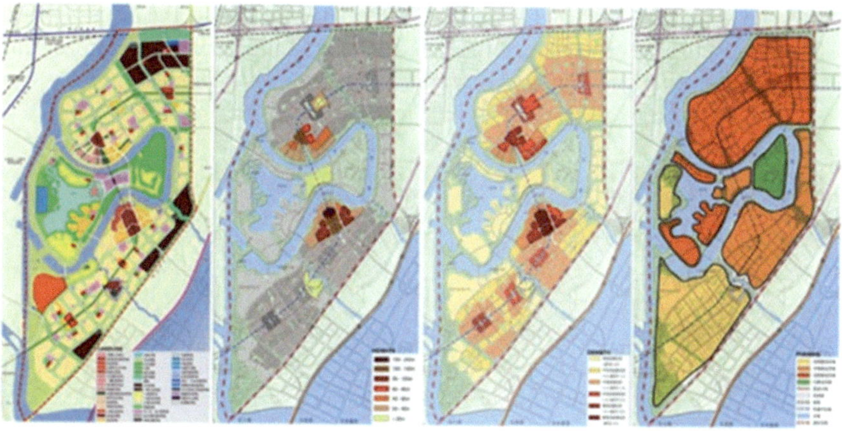

Fig. 13.5 Tianjin master plan (Source: Tianjin Eco-City/Surbana Urban Planning Group (2011). Retrieved May 28, 2016, from http://www.archdaily.com/102887/tianjin-eco-city-surbana-urban-planning-group/)

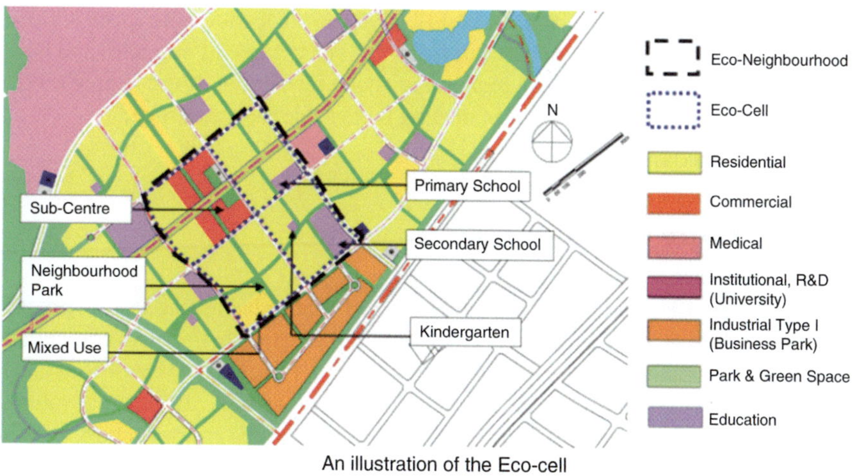

Fig. 13.6 The concept of cell (2010, January 14). Retrieved May 20, 2015, from http://www.cityup.org/ae_ecocity/Cases/Asia/20100114/58661-6.shtml

13.3.1 Health Benefits

Active travel improves physical, mental and emotional health (commuter challenge 2002) of people. By promoting car-free environment, one can eradicate or minimize to a great extent the air and noise pollution as well as congestion as observed in the case of Tianjin city. Issues like asthma, lung infections and heart problems can be reduced. Car-free streets and squares can become spots for children to play and encourage people to socialize by creating attractive environment and ensuring their safety like in the case of Vauban. Accidents and road fatalities can be minimized and a safe environment can be created for senior citizens and handicapped.

13.3.2 Economic Benefits

A considerable amount of money can be saved every year by avoiding the use of motorized vehicles both at individual level (petrol and maintenance cost of vehicles, tax) and government level (provision and maintenance of infrastructure). Public transport can provide an affordable solution with less subsidy demand for operating costs. Similar situation can be observed in the case of Tianjin city.

13.3.3 Societal Benefits

The settlement of Vauban as well as Bad Ischl suggest that with more space allotted to green areas, public parks, pedestrian pathways that are free of obstacles and have high aesthetic quality, there is an increase in social interaction. More people prefer to use public and community spaces making them safer and crime-free.

13.3.4 Environmental Benefits

All the three cases studied in this paper suggest that reducing the use of private vehicles and using public transport can result into creating better environment by reducing the pollution. By creating the green corridors like in the case of Bad Ischl one can conserve environment and also provide space for people to enjoy the nature. Also with lesser space utilized in building transport related infrastructure, it will become possible to conserve natural habitat.

13.4 Conclusion

The concept of eco-city aims to amalgamate environmental concerns and balance the growth of a city within the scope of carrying capacity of ecological system (Unescap). Out of a number of visions listed for the development of eco-city, many of them indicate that transport plays a key role in making a city truly sustainable. In variety of models of eco-city, even the ones that are not discussed in the paper, like Masdar City, Curitiba, Barcelona, Gyor, Tampere, etc., transportation plays a strategic role. Besides improving the quality of life, active travel and public transport also contribute to the economy. It is realized that by designing a compact city and a city with greater accessibility, one can reduce the amount of money spent on developing infrastructure to cater to motorized vehicles. It is also realized that a well-designed and efficient transport system, which gives priority to pedestrians and cyclists, contributes to economic growth and helps in refining the quality of life of the people, thereby becoming one of the most important sectors of urban development in the concept of eco-city.

References

Barton H et al (2010) Shaping neighbourhoods for local health and global sustainability, 2nd edn. Routledge, London

Black W (2010) Sustainable transportation. Guilford Press, New York, Print

Broaddus A (2010) A tale of two Ecosuburbs in Freiburg, Germany: encouraging transit and bicycle use by restricting parking provision. Transportation research of record, Vol. 2187, pp 114–122 Washington, DC

Commuter Challenge (2002) Additional benefits of sustainable transportation [online] Available at: http://commuterchallengebc.ca/additional-benefits-of-sustainable-transportation/. Accessed 17 Mar 2015

Coplak J et al (2003) Planning sustainable settlements. Publishing House of Slovak University of Technology, Bratislava [ISBN 80-227-1941-2]

ECOCITY, final report (2001) Urban development towards appropriate structures for sustainable transport. [online] Available at: http://www.rma.at/sites/new.rma.at/files/ECOCITY%20%20_%20Final%20Report.pdf chapter 4, pp 41–48. Accessed 12 Mar 2015

Mondal P Urbanization: causes and effects of urbanization in India [available at]-http://www.yourarticlelibrary.com/india-2/urbanization-causes-and-effects-of-urbanization-in-india-766-words/4659/

Skala F et al (un-known) Encouraging walking, the role of urban design experiences of the EU-ECOCITY Project. *Walk 21.* [online] Available at http://www.walk21.com/papers/Schubert,%20Uwe%20and%20Franz%20Scala-%20Encouraging%20walking,%20the%20role.pdf. Accessed 20 Mar 2015

Spotts, M (2013) Issue background: sustainable communities & transit-oriented development–[online] Available at http://www.enterprisecommunity.com/policy-and-advocacy/issues/green-and-sustainable-housing-and-communities/sustainable-communities-initiative#sthash.DqSf3vwC.dpuf. Accessed 20 Aug 2015

UNEP south– south cooperative case study (2013) The Sino-Singapore Tianjin eco-city: a practical model for sustainable development .[online] Available at http://www.unep.org/chinese/south-south-cooperation/case/casefiles.aspx?csno=114. Accessed 12 Mar 2015

Unescap (unknown) Low carbon green growth roadmap for Asia and the Pacific fact sheet- ecocity.[online] Available at: http://www.unescap.org/sites/default/files/17.%20FS-Eco-city.pdf. Accessed 15 Mar 2015

Urbanisation and sustainable mobility in India– [available at] http://www.theguardian.com/sustainable-business/india-sustainable-urbanisation-transport

Chapter 14
Bansal Haveli at Bathinda: Sustainability Paradigm

Bhupinder Pal Singh Dhot, Harpreet Sandhu, and Gaurav Jindal

Abstract India has divergent climatic zones. This situation has always posed challenges in the construction of buildings according to the typical climate of different regions. Traditional buildings by and large followed sustainable approach through relevant design strategies, whereas most of the modern buildings are found unable to adapt to the extremes of climate and are inherently energy consuming. *Bansal Haveli* at Bathinda in the Malwa region of Punjab (India) has been discovered as one of the best examples of traditional residential architecture. It truly conforms to the concept of sustainability as it has been evolved in response to the lifestyles and climatic conditions of the region and by using local materials. Thermal comfort has been achieved through design strategies like internal courtyards, orientation, thermal mass, and solar gain. Courtyards provide sufficient measure of comfort to its inhabitants by means of passive cooling system. Most of the rooms are directly connected to its courtyards which act as a transition space between the interior and exterior. No mechanical device is being used in the interiors to heat up or to cool the Haveli. Air flow and shading are the two components which help in regulating the temperature, keeping it warm in winter and cool in summer. Objective of the present research paper was to analyze the built form and characteristic features of the Haveli. It was found environment friendly, energy efficient, and gifted with spaces in harmony with the criteria of sustainability.

In this research paper, sustainability has been taken to mean the climate responsiveness of Bansal Haveli serving the aim of comfortable living.

Keywords Haveli • Courtyard • Environment friendly • Sustainable • Thermal comfort

B.P.S. Dhot (✉)
P.T.U. Jalandhar, Kapurthala, India
e-mail: bpaldhot@hotmail.com

H. Sandhu • G. Jindal
G.Z.S.P.T.U. Campus, Bathinda, India
e-mail: harpreet.sandhu_22@yahoo.com; gj2708@gmail.com

F. Seta et al. (eds.), *Understanding Built Environment*, Springer Transactions in Civil and Environmental Engineering, DOI 10.1007/978-981-10-2138-1_14

14.1 Introduction

Traditional buildings make substantial contribution to the quality of life by providing sensitivity about the past and by adding visual interest to the environment. Traditional architecture stood for construction work on strong and rich foundation of experience. People built their houses in harmony with the environment as well as with maximum utilization of the local materials. In the same context, the architectural built form of *Havelis* was evolved in response to climate and lifestyle using local materials (Slesin et al. 1998). The use of natural and passive means in Havelis proved effective in providing thermally comfortable space relatively warm in winter and relatively cools in summer. No mechanical device whatsoever was used to achieve desirable thermal comfort conditions (Khan et al. 2010). This feat was the sheer magic of architectural insights into planning, designing, suitability of locally available materials, and appropriate construction methods. In contrast, modern buildings are largely dependent on the use of mechanical means of energy due to their nature of built (Mohamad 2007).

The Western Punjab posed problems of comfort living because of its hot and arid climatic conditions. To meet the challenge of this harshness, the people of this region had to resort to the construction of environment-friendly buildings. Some of these residential buildings emerged in the form of Havelis (Randhawa 1999). Out of these, one example is *Bansal Haveli at Bathinda* – Punjab (India) – which was planned on pragmatic lines for ensuring comfortable living conditions. Comfortable environment was ensured through climatic modifying features such as internal courtyards, thermal mass, and solar gain.

14.2 Climate of Bathinda

Bathinda's climate corresponds to semi arid with high variations between summer and winter temperatures. Average annual rainfall is in the range of 20–40 mm. In summers temperature may go up to 48 °C and in winters temperature is as low as 1 °C. The weather is generally dry but is very humid from mid-July to the end of August. Rainfall is primarily from the south–west due to monsoon weather and is concentrated in the period from July to mid-September (Wikipedia) (Fig. 14.1).

14.3 Haveli

A clear definition of Haveli is not available, but its origins are traced to "Haowala," meaning partition in old Arabic. It is related to the term "Hawaleh" meaning "all-round" or "roundabout." It appears as if the Mughals used this term for defining a piece of land like the English term "estate" (Jain 2004). As for a physical definition,

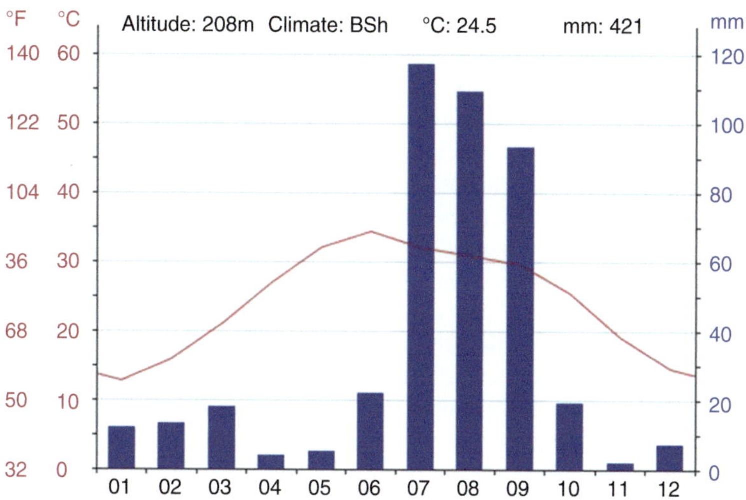

Fig. 14.1 Graph showing variation in temperature

it can be said that Haveli is a medieval dwelling with a courtyard or a bigger mansion within. The Havelis of Northern and North-Western India are perhaps the most popular types of courtyard houses in India and particularly in its states of Gujarat, Rajasthan, and Punjab (Saraswat 2011). The Havelis in Punjab reflect the psychology of the people who built them robustly relying primarily on security for safeguarding their families from the Muslim invaders. Typically the western section of Havelis is more open to the exterior and therefore is predominantly man area or *mardana*, while the eastern section which is carefully set away from the entrance is the women or *zenana* area (Randhawa 1999). The number of courtyards varies from a single one to more one. In most of the cases, open verandah lies in the front of the Haveli facing the street/road. There is usually one baithak (drawing room), occupied by males and used for gathering and entertaining guests. The features like doors and windows are intricately ornamented. Main entrance door is usually fabricated from solid wood and is divided into rectangular planes or engraved panels. Rainwater is also stored underneath the courtyard for various uses.

Features reflecting the sustainable character of the Haveli are as under.

14.3.1 Spatial Organization

The Haveli is located in the central core of Bathinda city area popularly known as Sirki Bazar. Its region has a typical desert climate where during the summer seasons temperatures are as high as 48 °C and during winters as low as 1 °C. It was built about 100 years ago and has three-storied structure. The main feature of its spatial organization is symmetrical planning worked out around two courtyards which delineate public and private spaces. The front area is used for public gathering and by the males, whereas the rear area is for privacy and is meant for females of the family as per Indian traditions. Courtyards and terraces on different levels are most fascinating features of this Haveli which not only define the perfect spatial organization but are also the center for rituals, marriages, festivals, and social gatherings. Courtyards, nearly square in shape give connectivity to areas or spaces inside and outside. The vertical part encloses courtyards and defines their volume. The elements like arches, niches, fireplaces, and cornices make this space lively.

Well-placed and well-proportioned courtyards at all levels also serve as microclimate modifiers due to their ability to mitigate high temperatures and to channel breezing and thereby adjusting the degree of humidity. During summer time, throughout the day, they provide shade and in combination with the thick external walls delay the heat gain. During night, warm air rises and exits from the courts. The cool air enters to supplant already existing air. Hence, during hot day, cool air gets circulated to the rooms and the courtyards turn into a source of fresh and cool air. These aspects clearly reflect sustainable character of Haveli (Figs. 14.2, 14.3, 14.4 and 14.5).

14.3.2 Room Size and Massive Walls

In the Haveli, majority of the inner rooms on the ground and on the first floor face inwards and are dependent on the courtyards for daylight. Longer sides of the rooms lie along the courtyard and their depth is comparatively shallow. The walls of this Haveli are massive with thickness varying from 18″ to 27″ and were constructed with Nanakshahi bricks, mortar of lime and surkhi. Thermally heavy weight construction is part of the climate responsive strategy for both the cool and the warm periods. Furthermore, thick walls in addition to their insulating properties, act as heat reservoir. During hot days, the heat that flows from exterior due to solar radiation to the inside gets reduced. During the cool hours, a part of the stored heat in the walls is released to the interior. This results in the minimization of temperature change inside the Haveli. On the other hand, in winter, heating requirements are reduced due to the heat stored in the walls which radiates during the night.

14.3.3 Verandahs, Overhangs, and Balconies

Verandahs are located at the entrance and around the courtyard, shading the peripheral rooms. They function as a transitional space between enclosed rooms and outdoor spaces and also provide shade to the walls reducing heat gain. Its verandah, on the southern side, is the most effective building element which allows the sun to reach the interiors in winters and prevents it in summers (Fig. 14.6).

Fig. 14.2 View of rear courtyard

Fig. 14.3 View of front courtyard

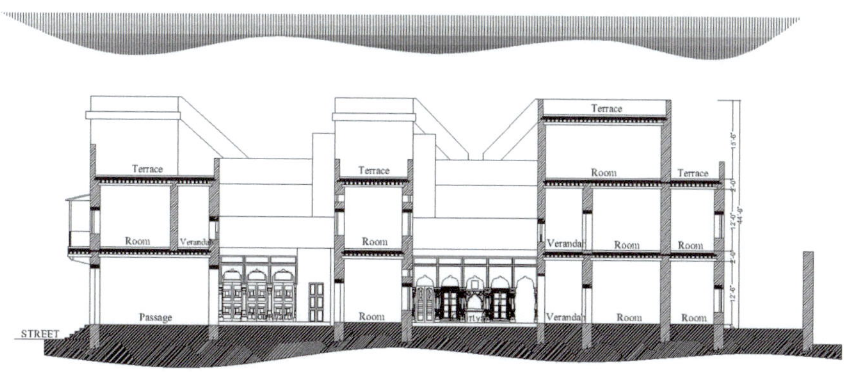

Fig. 14.4 Section showing built and open space relationship

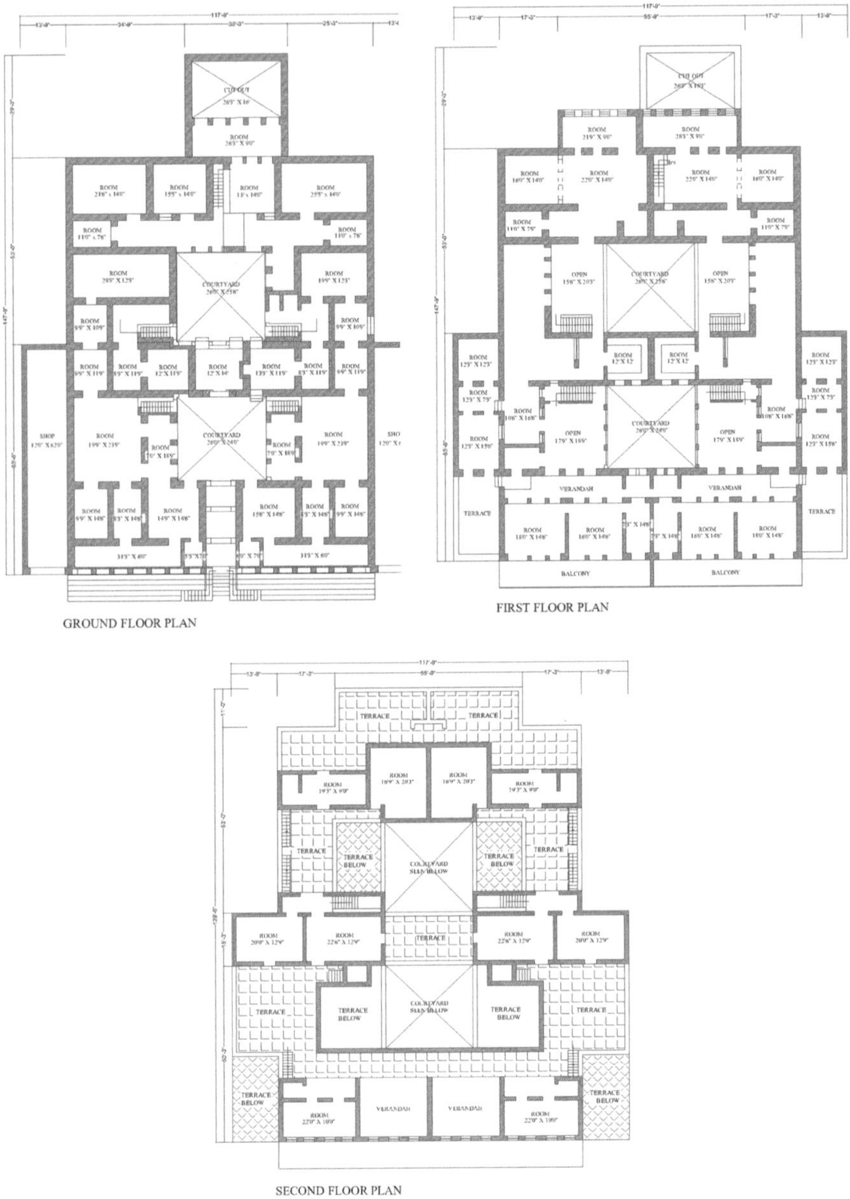

Fig. 14.5 Floor plans at different levels

Fig. 14.6 Verandahs on southern side for reducing solar radiations

Fig. 14.7 Covered passage to connect two courtyards and balcony above opening

Projections at different levels shield the Haveli from direct solar radiations and reduce heat gain. The depth of balconies and overhangs minimizes the solar radiations in summer and allow the winter sun to come therein. Balconies have been projected in front of large openings to prevent solar radiation from entering the rooms through them whereas balconies on the first floor have been blocked by wooden jaali windows to create more privacy from the street side (Fig. 14.7).

Fig. 14.8 Recessed openings provide sufficient glare-free daylight and ventilators near ceiling create stack effect

14.3.4 Heavy Roofs, High Ceilings

Due to orientation and mutual shading, the walls stand protected from solar radiations. The thickness of roof varies from 15″ to 18″. The flat roof is made of timber. Timber being bad conductor of heat does not allow the horizontal surface to gain much heat throughout the day. Greater ceiling heights of 14–18′ improve environmental conditions in summer time by permitting warm air to rise. These heights increase the volume of the enclosed space, taking thus more time for the internal air to get heated up as compared to the buildings with low height ceilings (Fig. 14.8).

14.3.5 Openings and Ventilators

Ensuring air movement in the built space through the openings provides thermal comfort at higher temperature. These are efficiently shaded from direct solar radiation. However, in winter when there is no special need for air movement, window apertures are opened during the day to store the thermal radiation, and these are kept shut at night. Windows also provide sufficient daylight into the interior. Ventilators are prominent features in this Haveli. These are manually operated and stand provided just below the ceiling. The warm air rises and leaves the space. The cool air from the courtyard enters into rooms, thus creating the stack effect. A typical ventilator near the ceiling increases the velocity of air entering into the Haveli and results in lowering of the pressure at the ceiling level, thereby inducing the hot air under the roof to flow out. Ventilators also function as clearstory windows to light up the interior spaces not having any exposed surface to admit light through window openings.

14.3.6 Surface Texture and Color

The textured surfaces are used in the exterior finish of the Haveli facades exposed to sun. The walls are finished with lime and stucco plaster. The façade of the Haveli is treated with stucco on motifs and floral patterns made up of lime plaster. The external surface of the Haveli is painted with light colors that reflect solar radiation in order to have minimum absorption. Whitewashing reduces the absorptive property of the wall surface, minimizing the effect of solar radiation on internal climate and tends to stabilize the inside temperature.

14.4 Observations

Going by the perspective of climatic responsiveness, the principles of planning and designing were precisely worked out and actually followed. Natural and passive means has been employed for reduction of dependence on mechanical energy. Thermal comfort has been achieved through design features like internal courtyards, orientation, a thermal mass, and solar gain. It highlights the construction methods and skills in the use of locally available materials.

These observations would be theoretically and practically useful to the professionals working in the field of climatic responsive architecture.

14.5 The Conclusion

Overall planning of the Haveli with two courtyards, its construction with locally available materials, its immunity from the harshness of summer–winter seasons, and its naturally controlled flow of air and openness for natural light reducing dependence upon mechanical energy, constitutes cogent proofs of its sustainability paradigm. This kind of combination of the various features provides comfortable environment to the inhabitants. A variety of spaces like open, semi-open, and enclosed is being used during daytime in different seasons. The courtyards ensure ventilation in the building. Verandahs on the outside facing balconies covered with wooden windows help in reducing the heat intake from south–west side. These serve as buffer space between the interiors and the outside environment. The walls well protected from the sun with various elements cool the building interior. Building components and spatial layering in the Haveli help in the reduction of heat intake. The greater ceiling height increases the volume of the enclosed space, thus taking more time in heating up the internal air. There is time lack due to thick masonry walls and heavy roof construction system. The openings such as windows and ventilators provide cross ventilation which helps in creating stack effect. Building mass reduces the heat gain in summers. Its harmony with the environment both physical and social along with its solidity and intactness clearly discernible even after 100 years of construction are also significant factors constituting its sustainability.

Obviously this Haveli has significant lessons for modern-day construction works of diverse types. It is hoped that the outcomes of this study would provide necessary insights to architects for achieving the goals of energy-efficient and sustainable design solutions in their professional ventures. Traditional buildings are not just remnants of the past but great lessons for the future of mankind. So the need for making present-day built environment sustainable may well be served by using climate responsive designs grounded in traditional architecture.

References

"Climate of Bathinda" Wikipedia

Khan MS, Sethi Haveli (2010) An indigenous model for 21st century Green Architecture. Archnet-IJAR 4(1), from website www.archnet.org/library/documents

Mohamad AK (2007) Climatic responsiveness in traditional built form of lucknow, Ph.d. thesis

Randhawa TS, (1999) "The Indian Courtyard House"

Saraswat S, (2011) "Understanding courtyard design through Haveli's"

Shikha J, (2004) "Haveli's: A living tradition of Rajasthan"

Slesin S, Cliff S (1998) "Indian style". by Thames & Hudson Ltd (July 16, 1990)

Other References

Dili a S, Naseer MA, Zacharia Varghese t (2010) The influence of internal courtyard of kerala traditional residential buildings in providing a comfortable indoor environment

Dr. Mahmoud Ahmed Eissa, "Ecological Aspects of the Courtyard House as a Passive Cooling System"

Gautam, A (2008) "Climate responsive vernacular architecture: jharkhand, India". (online) Available at http://krex.k-state.edu/dspace/handle/2097/990

Part II
Urban Regeneration and Sustainability

Chapter 15
Politics of Water and Development: Case of Pune

Sharduli Joshi

Abstract Since early civilizations, rivers and streams are modified by human activities. These rivers and streams nurtured the civilizations to become urban areas. But these urban areas have turned their backs to the urban streams and rivers. It is crucial to better understand and integrate its functioning to urbanization dynamics.

If we narrow down, mismanagement of water resources which has given rise to the twenty-first century slogan that "Nothing is holy, sacred, or off limits when everything is for sale." This mentality leads us to mercilessly pollute the Earth, discharging industrial waste and sewage water into rivers and lakes.

A 2006 United Nations report stated that "There is enough water for everyone," but that access to it is hampered by mismanagement and corruption (UNWATER, Water, a shared responsibility, the United Nations World Water Development Report 2. UNESCO, 2006).

Today the struggle for scarce water resources in many places is unavoidable. Many river basins are unable to fulfill the demands of water even for their rivers to reach the sea. Further extraction of water for human use is not possible because limits have been reached and in many cases breached. Greater competition raises questions between states, countries, and regions over allocation of water, which is benefited, between those upstream and those downstream.

As politics (the process of decision-making of groups of people, involving the authoritative allocation of, e.g., resources), the actors, their interests, and interactions determine whether progress is made or hindered, it is important to understand the politics of water, and how it impacts on development is the key to improve water and development scenarios.

The paper analyzes the case of politics of water in the city of Pune and the surroundings. It discusses the issues of environmental flow, channelization, and several other factors related to development. It concludes with suggesting policy recommendations for afforestation, recycling of waste water, rainwater harvesting, etc. to resolve the issue of water shortage for environmental purposes.

S. Joshi (✉)
Brick School of Architecture, Pune, Maharashtra, India
e-mail: shardulijoshi@brick.edu.in

F. Seta et al. (eds.), *Understanding Built Environment*, Springer Transactions in Civil and Environmental Engineering, DOI 10.1007/978-981-10-2138-1_15

Keywords Urbanization dynamics • Mismanagement • Over allocation • Politics • Development

15.1 Introduction

Water is probably the only natural resource which is a primary requirement in all aspects of human civilization from agricultural and industrial development to cultural and religious values embedded in society. The need and demand for water has been a driving force for health, society, economic prosperity, cultural significance, and development throughout human history.

The city of Pune was established around 1,000 years ago near the confluence of rivers Mula and Mutha on the western side which was considered sacred to build any settlement due to its physical and religious significance. The river Mutha has been the structuring element of the city which divides the old and the new development.

There are five major rivers flowing through Pune city and nearby industrial area. Mula River is dammed thrice before it reaches to the city first at Panshet and then at Khadakwasla and Temghar.

Mr. Rajendra Singh, known as "waterman of India," quotes that the conditions of Mula and Mutha rivers passing through the city are worse than that of the Ganga (TOI 2013).

In research and site visits for this paper, it was observed that Mutha and Ram rivers collect wastewater from the residential areas of the city. Mula and Pavana rivers are mostly contaminated by industrial discharges, agricultural runoffs, and sewage. Indrayani River encircles the city passing through Alandi which receives wastewater from newly emerging industrial and residential areas from Dehu to Alandi. Pavana joins Mula River near Dapodi, while Mutha joins Mula River near the College of Engineering, Sangam Bridge. The combined river Mula crosses down south and joins Bhima River as shown in Fig. 15.1.

All these rivers are the victims of ultramodern high-tech development in their catchment areas (Joshi 2015).

15.2 Urban Stressors

15.2.1 Increasing Population and Urban Waste

The population growth of Pune is more than 0.2 million per annum which burdens the drainage systems leading to filthy living conditions with improper sewage disposal (Joshi 2015).

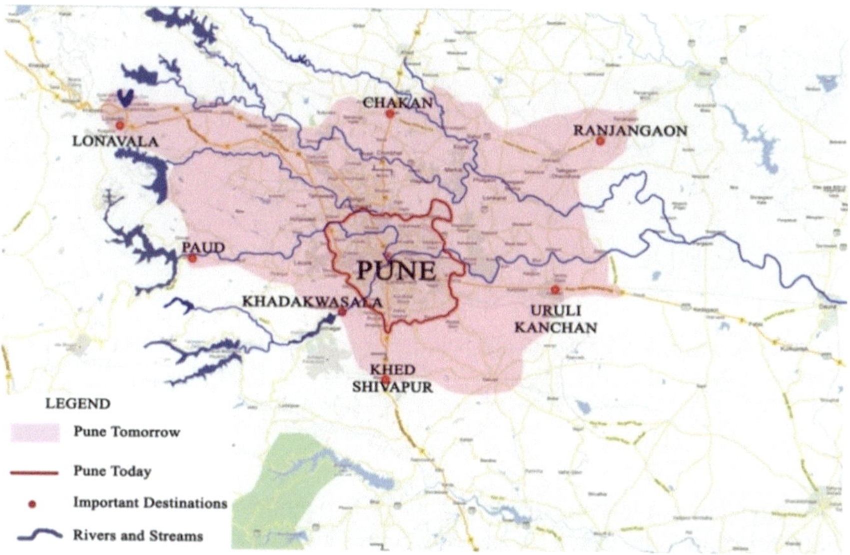

Fig. 15.1 Pune and its watershed (Source: Prepared for academic purpose with the help of Google maps)

Table 15.1 Values of DO, BOD, and COD of survey against values in ESR (2010)

Name of the point	DO (ESR)	DO (survey)	BOD (ESR)	BOD (survey)	COD (ESR)	COD (survey)
Mhatre bridge	1	0.6	30	140	55	320
Sangam bridge	1.2	0.5	26	100	80	360

(Source: Survey findings and ESR report, Pune 2010–2011)

A survey was carried out in May 2013 to check the water quality in terms of DO, BOD, and COD of the river at two different points. The results were drastically different from ESR, Pune 2010–2011 (Environmental Status Report). Refer to Table 15.1 for exact figures (Fig. 15.2).

Also in one of the interviews for TOI, Rajendra Singh states that "There have been attempts to get funds from the Central and state governments for restoration of these rivers. However, it seems that nothing specific has been done. We don't see serious steps by the administration in conserving rivers" (TOI 2013).

Inadequate sewerage system for four million strong population in Pune Metropolitan Area is leading to the pollution of natural drains in the city's watershed. These rivers have become large gutters carrying the sewage from the city.

Meanwhile, pollution-control authorities, government officers, journalists, citizens, and policy makers of state and central governments are expressing the urgent need for affordable indigenous pollution-control techniques which will improve the quality of environment for the society (Jaymala Diddee 2000). There is a gross system failure as the city service providers are not able to keep pace with the city's growth.

Fig. 15.2 Survey done for water quality testing at approximately the same points as ESR with reference to Table 15.1 (Source: Photographs clicked during the survey)

The pollution is reaching the water bodies through natural drains and streams that are used for drinking and irrigation purposes, making them unfit for any application. The contamination of rivers by sewage and industrial effluents has led to eutrophic conditions and dense growth of a particular aquatic species like water hyacinth and also caused the blackening of water due to accumulation of organic matter (Roy 1973). Thick mats of water hyacinth reduce the velocity of the river, thus becoming a breeding ground for mosquitoes and other nuisance insects. It adds to the problems downstream where the deposited organic matter undergoes anaerobic degradation.

15.2.2 Environmental Flow

Humans are not the only ones who are using available water for themselves. Aquatic, terrestrial ecosystems are also dependent on fresh water. Increasing economic activities as well as the standard of living is affecting the quality and quantity of water in the rivers. While deciding environmental flow to any river basin, usually two questions arise: (1) How to define environmental water needs? (2) How to quantify them? (Venot 2008). The climate is changing, affecting every aspect of neighborhoods, economies, and ecological systems.

If environmental flow is considered as a sector along with other sectors like irrigation, industries, etc. which compete each other for water, it is seen that the sectors other than environment have a voice, and they can pay for the services. Environmental sector neither has voice nor can it pay for its water demand (Upali 2009). Though 5 % water of the storage capacity is allocated for environmental flows, it is often neglected and is consumed by one or the other sector. Figure 15.3 shows the water flow in Bhīma subbasin in monsoon and post-monsoon months, respectively.

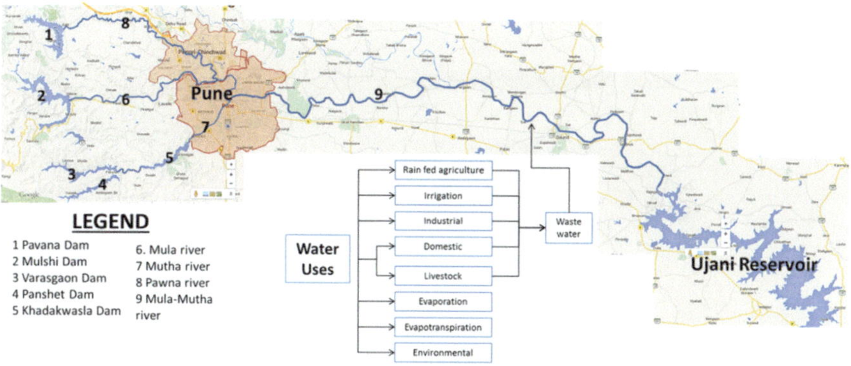

Fig. 15.3 Water flow in the river in monsoon and post-monsoon months, respectively (Source: Prepared for academic purpose)

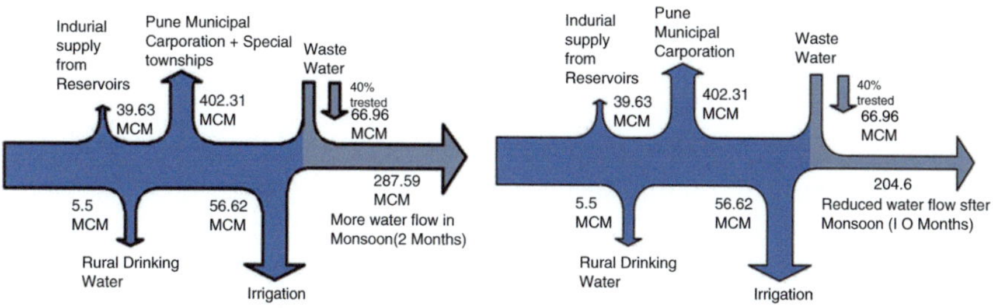

Fig. 15.4 Flow path of urban streams of Pune (Source: Prepared for academic purpose)

15.2.3 Water Development

Pune is facing acute shortage of water since the last few years. In the month of April every year, the region is facing water crisis. Pune has five rivers (shown in Fig. 15.4) and four dams that supply water, which are more than enough to meet the city's needs. A lot of water is unfortunately wasted by citizens and farmers. Also, there is unequal distribution of water between the city and neighboring agriculture requirements.

The Mula-Mutha River has been dammed thrice upstream to provide water supply to the city, Khadakwasla, Panshet, and Varasgaon being the locations. The reason for damming the river is to hold water for certain areas for irrigation and not leave it ahead for irrigation. All of this is part of politics controlled by some influential people.

Maharashtra has the highest number of large dams in the country and is now claimed to be suffering the worst drought in four decades or more.

Hindu (2013) writes that at the post-budget press conference, Deputy Chief Minister Ajit Pawar conceded that 70 % of the water went to sugarcane cultivation

Table 15.2 Water uses in the study area for different purposes (unequal and non-priority distribution of water)

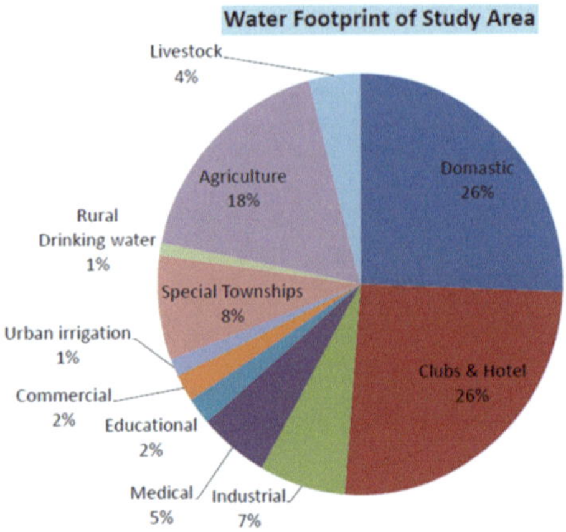

Water footprint of Study Area	
Domastic	146.65
Clubs & Hotel	146.65
Industrial	39.63
Medical	31.71
Educational	11.89
Commercial	11.89
Urban irrigation	7.93
Special Townships	45.60
Rural Drinking water	5.50
Agriculture	101.26
Livestock	23.36
	572.06

(Source: Prepared for academic purpose)

in rural Maharashtra. It further mentions, "As on 31st December, 2012, out of the total sugar production in the country, the share of the Maharashtra was 35.3 per cent." So the drought-prone districts produce more than a quarter of India's sugar. Table 15.2 shows unequal and non-priority distribution of water in Pune.

It slams the government for not making any attempt to curb either planting of sugarcane or other water-intensive crops or to curb any of the water-intensive activities like running of sugar and wine factories in drought-affected districts.

SANDRP (How is 2012–2013 Maharashtra drought worse than one in 1972? 2013) blames GoM, for making wrong decisions about building impracticable and unwanted dams, cropping patterns, unacceptable water management system, diversion of water for non-priority uses leading to reduction in per capita water.

15.2.4 *Efforts of Restoration*

A river restoration and navigation project is being planned in Pune, and before it is formally accepted, it seems that it is receiving special favors from the Pune Municipal Corporation (Yadvadkar 2011). Pune massively pollutes its rivers, with more than 744 million liters of sewage generated per day, out of which barely 30 % is treated. Rather than focusing strongly on river restoration and pollution control, this project, which is said to receive funds earmarked by JNNURM for river restoration, is focusing more on navigation. The works, which include large-scale channelization of the river, seem to be at cross purposes with restoration (Yadvadkar 2011).

The state water policy was formed by GoM in 2003 and was revised in 2011. The policy describes the water scenario, challenges, and perspectives. Policy framework emphasizes on Integrated Water Resource Management (IWRM) along with the need for watershed management, groundwater management, and aquifer management. The revised policy gives second priority to irrigation after drinking water (Patil 2012). Water for environmental purposes is still at fifth place. The policy mentions that dams should allow the minimum water flow as environmental flow in the rivers, but the minimum water flow is not defined as well as the water rights are not clearly mentioned. The political economy operates the water market in Maharashtra. If the water policy is rightly drafted and followed, some of the answers to water scarcity and droughts in Maharashtra can be fetched.

The state of Maharashtra has established Maharashtra Water Resources Regulatory Authority (MWRRA), in 2005 to finalize water tariff for irrigation and non-irrigational water uses with appropriate incentives and penalties. The establishment of such a body is quite creditable as it is the first state in India to form such a body which is responsible for monitoring and regulating water resources within the state as well as its trading, allocation, and utilization to maximize efficiency of water use.

15.2.5 *Channelization of the River*

There are plans of beautifying Mutha River under the river restoration plan of JNNURM which supports the riverside road project (IndianExpress 2011). This is an inspiration from the so-called successful Sabarmati Riverfront Development project which has become a hit and is being looked at as a model project for developing the riverfront in any city without giving a closer look to the local needs and context (ITDP and Embarq 2012). Several NGOs in the city pointed out that the detailed project report (DPR) does not really focus on restoration of the river, but on preparing a backdoor entry for river navigation and real estate development (TOI, River restoration plan draws flak 2011).

"The road construction in the river is violation of several environment norms. In such case, there is no other space for the road. So, instead of dumping mud and rubble, the authorities should construct a raised platform for the road which would allow free flow of water and avoid flooding of the area" (Yadvadkar 2011). To elaborate more on this, many nongovernmental organizations (NGOs) are opposing this idea of the road as it sits within the riverbed. Any construction within the riverbed is against the River Act of 1958. This would put an end to the natural condition of the river and make the environment more engineered with channelized hard edges (Fig. 15.5).

At present, instead of making new policies and governing bodies, collaborative and synchronized working of existing bodies is important. Also the grassroot level actions in individual subbasins and water sheds should lead to basin-level priorities.

Fig. 15.5 Channelization of Mutha River in Pune (Source: Clicked for personal use)

In the Bhima subbasin, MWRRA is facing the opposition from downstream users because of the very poor quality of water in the "Ujjani" reservoir because of untreated sewage from Pune and suburban areas. As this upstream area is a sugarcane belt, downstream cities like Solapur are experiencing scarcity of water. The dams located upstream have huge amount of dead storages.

During the survey conducted by SANDRP, it was observed that few hotel managers denied that there is water scarcity in Solapur. This shows that the poor are suffering from water scarcity, whereas the wealthy and powerful continue to get water whenever they want, regardless of the quantity. Pune ESR shows that 146 MCM water is diverted to clubhouses and hotels in urban and suburban areas of Pune which is equal to domestic water requirement of the entire city.

15.3 Conclusions

Water accounting and auditing is extremely important in order to save water by reducing water losses in undesirable sectors and for creating transparency and accountability within the system.

The political economy operates the water market in Maharashtra. If water policy is rightly drafted and followed, then some of the answers to water scarcity and droughts in Maharashtra can be fetched.

Some of the strategies to resolve the water crisis are afforestation to reduce runoff and salinity to increase availability of water, recycling of waste water which will reduce demand of freshwater, change in irrigation pattern and use of recycled water, rainwater harvesting in urban areas to increase availability of water, and storage of water through precipitation. These strategies are essential to make a change because "water is life"; otherwise water would be the cause of a third world war.

References

Hindu T (2013, April 4) Maharashtra drought man-made, says study. The Hindu

IndianExpress (2011, December 17) River plan hits a hurdle over cost break up. Indian Express

ITDP, Embarq (2012) Life and death of urban highways

Jaymala Diddee GS (2000) Pune: queen of Deccan. Eliphant design Pvt. Ltd., Pune

Joshi S (2015) Streams of Pune: waiting for restoration. In: Joshi S (ed) River me and environment. Menaka Prakashan, Pune, p 165

Patil (2012) Reforms in water resources sector in Maharashtra State, India. India Water Week 2012 – Water, Energy and Food Security, Water, Energy and Food Security (April 2012), pp 10–14

Roy M (1973) Rivers in the city. David Charles, Newton Abbot

SANDRP (2013) How is 2012–13 Maharashtra drought worse than one in 1972? SANDRP

TOI (2011, November 7) River restoration plan draws flak. Times of India

TOI (2013, May 1) Civic bodies must draft policy to restore rivers of Pune: expert. Times of India

UNWATER (2006) Water, a shared responsibility, the United Nations World Water Development report 2. UNESCO

Upali AT (2009) Strategic analyses of the National River Linking Project (NRLP) of India series 1. IWMI, New Delhi

Venot SB (2008) The lower Krishna basin trajectory: relationships between basin development and downstream environmental degradation. IWMI, Colombo

Yadvadkar S (2011, March 22) Fate of Pune's rivers: presentation by Sarang Yadvadkar. Retrieved March 2015, from India Water Portal: http://www.indiawaterportal.org/articles/fate-punes-rivers-presentation-sarang-yadvadkar

Chapter 16
The Implementation of *Phulkari* Embroidery Pattern in Interior Decoration

Rajinder Kaur and Ila Gupta

Abstract In architecture, decoration or ornamentation is an art, used to embellish home and interiors which constructs a pleasant and peaceful ambience in the surrounding through aesthetic, colorful, and creative artifacts. Art has a remarkable significance in the production of interior design because every concept has been associated with the features of beauty and ornamentation in interior decoration. The term decoration is particularly associated with art and its fundaments of beauty, harmony, balance, color application, symmetry, and so on. In this way, art plays a magnificent role to decorate the surface with comprehensive attributes. Interior decoration has numerous elements like furniture, decorative items, wall decoration, and many more. Textile is one of these elements. Textile is, the most important tool, extensively used in interior accessories which includes bedsheets, carpets, cushion covers, wall hangings, and covers for furniture, walls, and ceilings. Today, embroidery is not only a technique of cloth embellishment but also a means for artists to express their inner feelings and emotions through the established aesthetic concepts. In contemporary society, embroidered textile is used for interior decoration items, to create a traditional and royal appearance. *Phulkari* (floral work) is a traditional craft which reveals at Punjabi (Punjab, India) women on the canvas of the cloth in specific geometrical motifs. They use a needle and thread with an unlimited color palette with skillful, artistic, and aesthetic appeal for the commercial purposes. The present paper focuses initially on the use of traditional handicraft *Phulkari* patterns in modern interior decoration like beddings, cushion covers, sofa covers, etc. The paper will also be an attempt to discuss the future perspective of the application of *Phulkari patterns* in interior ornaments with the special help of Photoshop software. Some designs of *Phulkari* embroidery have been taken as a study and placed on some blank interior objects after comparing both objects.

R. Kaur (✉)
Department of Humanities and Social Science, IIT Roorkee,
Roorkee 247667, Uttarakhand, India
e-mail: ahsasart.87@gmail.com

I. Gupta
Department of Architecture & Planning, Indian Institute of Technology Roorkee,
Roorkee 247667, Uttarakhand, India
e-mail: ilafap@gmail.com

F. Seta et al. (eds.), *Understanding Built Environment*, Springer Transactions in Civil and Environmental Engineering, DOI 10.1007/978-981-10-2138-1_16

Keywords Interior ornaments • Contemporary *Phulkari* • Punjabi women • Home • Motifs

16.1 Introduction

Embroidery is an art, which has been recognized as a heritage in almost, all over the world. It has always remained a fusion of old and new tradition and experiments in various styles and mediums and also with its all new dimensions of technology. Further, it is the biggest phenomenon in the fields of fashion (Lehri 2006). Embroidery is a beautiful art which evoke emotions and sentiments in the life of women. This art is prominently followed from generation to generation as a cultural heritage. The threads of embroidery are symbolic arranged to denote the patterns of the life cycle of woman (Paine 1990; Harvey 2002; Crill 1999). India is a land of multiple cultures; every tradition is blessed with its folk art which reflects the expression, interest, and talents. The folk art style varies from state to state. Punjab is situated in the northwest part of India, which is famous for its art and craft such as weaving, embroidering, mud wall painting, wood work, etc.; this particular art form has a rich cultural tradition in the state. Today, the uniqueness and richness of culture and tradition are beautifully kept alive and maintained in the government, museums, and NGOs, so that young generation will be aware of the richness of Punjab's folk art. Punjab is famous for its traditional folk art of *Phulkari* from the fifteenth century.

The women of Punjab were well trained, in embroidering *Phulkari* by their elders instead of education. *Phulkari* is a needlework of Punjab, which reflects the emotions and expressions of women through the specific embroidery technique. The word *Phulkari* is made from the Sanskrit word which means flower work done by women on the handmade, hand-spun cotton called "*Khaddar*." In the process, *Phulkari* is done with bright-colored silken threads called "*Pat*" mainly red, orange, blue, green, and so on. These colors are representative of Punjab's vibrant spirit and dynamic aspects, which can be easily seen throughout the Punjab. *Phulkari*, in traditional terms, has been used for covering the head of the woman and mainly associated with the *chunnis* or *dupattas* and also been used as a shawl to keep the body warm in winters. A simple embroidered *duppat*a or *chunni* with *butties* (little motifs) is used for daily wear known as *Phulkari*, and the kind of heavily embroidered *duppata* which is used to give and wear on special ceremonial occasions is called as *Bagh*. It is particularly used for women for some religious purposes or used to dress up for some auspicious ceremonies. The *Phulkari* is embroidered on the reverse side of the cloth, with geometrical and natural motifs by counting thread with darning stitch, by women of the place. It plays a major role in every occasion, for wedding, birth of a son or daughter, and death, which symbolize the happiness, prosperity, and at good end. Maternal grandmothers, of the region, take pains, care, attention, and pride of embroidering "*chop*" (a type of *Phulkari*). This, further, is used to give as a gift to their granddaughters on their wedding (Hitkari 1980). The art of *Phulkari* has demand and was famous not only in India but also in abroad too. Women create a design with their explicit imagination and creativity, without tracing or counting the stitches. In 1881, the first exhibition of *Phulkari* has been initiated in Punjab and gets recognition from numerous Westerns. This specific *Phulkari* has been started to embroider on purse, curtain, piano covers, and coats or tunics, as per the taste and demands of the Western market. Then, during the period of Maharaja Ranjit Singh (from 1881), it was exported to Western countries, and this exportation is being continued till today.

In contemporary aspect, creativity can be shown in numerous fields, including art, architecture, music, cinema, and so on. All these creative elements of representation have always been interrelated with each other with various sequences. This creative genius of Indian subcontinent has also been filled with traditional and cultural approaches. In recent time, interior designing is the most popular phenomena worldwide. Today, the purpose of interior design is not only limited in the dimensions of usability and reasonability, but it should also have aesthetic sense and artistic creativity. Furniture and other décor items create rich and high appearance, which give people more happiness. The aesthetic sense of people has been changed; now, it is growing day by day because everyone wants a particular look for their places. For instance, some people prefer traditional objects and some of them prefer royal one. In current trends, the aesthetical appeal has become an essential part of human life and it is related with their entire way of living. Thus, like dressing, living places or official places also come in the category of showing their particular taste or interest. They want to show their inner sense of aesthetic through certain things, and the interior decoration has also been done to show the same artistic sense. Then, in this way, the interior of a place has also become an important part in the field of architecture. In this context, people are going through their indigenous or native style to decorate their surfaces because after using such traditional arts, they not only decorate their surface but they also demonstrate their regional arts. Therefore, traditional art and craft pattern have been used in interior design with some new techniques and color in useable and decorative manner. These traditional designs have been highly demanded by consumers and decorators. In recent times, interior designing has been usually done in an ethnic and royal look with traditional art pattern. Traditional embroidery is one of the popular mediums which have been used in interior design and décor items. Traditional embroidery has a unique attraction owing to its patterns and colors, which give people a pleasurable ambience and change the whole environment of living space. *Phulkari* is not only restricted to be done on shawls or *duppta* (scarf) but also being done on scarves, saris, suits, file cover, mobile cover, clutch, bag, bed covers, home furnishing, etc. in bright and vivid colors. In the interiors of contemporary buildings, *Phulkari* has been taken as a medium of decoration which includes various handloom items too. These handloom items are cushion cover, bedsheet, table cloth, curtains, pillow covers, etc. In this context, *Phulkari* has also emerged as a medium to decorate the interior of dwelling places and to give empowerment for woman because it is particularly done by women. The creative handloom items are also used to cover the surface and keep the dust away from those utility items. This paper is an attempt to discuss the role of *Phulkari* in textile designing and interior decoration after analyzing its patterns in interior ornaments. In which, *Phulkari* is being emerged as a best component of decoration, having various approaches in itself. These approaches relate to the choices of individuals, either they are interior designers or consumers.

16.2 Methodology

This study is based on primary and secondary data. The primary data is collected from a survey of the market of Ludhiana, whereas the secondary data is taken from relevant books, journals, and articles. Some photographs have been taken from the "*Phulkari* Punjab government emporium." The methods, which are used, are descriptive and analytical. The major focus of the present study is to analyze new uses of *Phulkari* in contemporary terms on products and aim to promote and preserve Punjab's traditional handicraft. The paper also sheds light on some products which are used as interior decoration. Further, this paper also showcases some experiments on *Phulkari* patterns in interior decoration after using Photoshop software.

16.3 Review of Literature

Embroidery created by women is an expression of creativity which enrich their environment through creating objects from daily life (Dhamija 2004). There is no certain time frame, which gives the actual account of its origin, but during the fifteenth century, it is mentioned in the statement of the founder of Sikhism, Guru Nanak Dev (A.D. 1469–1538). According to him, "*Kadh Kasida Pehreh Choli, Tan Tum Janoh Nari*," which means you can be accepted as a woman, when you embroidered your own blouse with embroidery stitch (Hitkari 1980). *Phulkari* is playing a major role in the life of Punjabi women because it was used in every occasion related to the life of Punjabi people (Chattopadhyay 1975; Das 1992; Dongerkery 1951; Banerji 1955; Naik 1996; Dutta 1985). Little girls learned *Phulkari* technique from their grandmother and mother (Dhamija 2007). The upper-middle-class families give 50 *Phulkaris* and *Baghs* in dowry as a gift; it is a custom of Punjabi tradition (Pal 1955). Embroidery (*Phulkari*) was done on locally available cotton cloth in traditional Punjab by women (Himadri 1982). Cotton spinning and cloth weaving have commonly been done by the women of Punjab for personal and other domestic usage among both rural and urban Punjab (Tondon 1961). A soft thread of untwisted silk, which is called *pat*, was used for embroidery which is used to import from various places like Kashmir, Bengal Afghanistan, and Turkistan and then dyed locally, and silk thread was dyed in Amritsar (Punjab) and Jammu (Das 1992). The colors used in dyeing were made from leaves, flowers, barks, and roots. The process of making *Phulkari* was purely organic including dyeing of thread or strengthening of the fabric and entirely free from chemicals (Gillow and Barnard1991). "Eastern" Punjabi women embroidered *Phulkaris* with geometrical as well as figured designs including lozenges, squares, triangles, animals, plants, jewelry, and human figure; counting the thread women are also tracing the complicated designs in outline with black ink and then filling them in with darning stitch (Hitkari 1980).

Stylized vegetable and flowers like *karela* (bitter gourd), mustard flowers, golden yellow marigold, jasmine buds, lotus, and human images were embroidered with the technique of *Phulkari* (Dhamija 2007). Different types of vehicles as motifs like a railway train, motor car, and *Ratha* (chariot) had also been embroidered in *Phulkari* art by women. There are also some jewelry motifs like *hansali*, *tika*, *karda*, *ponchi*, *singar-patti*, and *guluband*, which were embroidered in traditional *Phulkari* (Aryan 1983). *Sainchi Phulkari* used to do with figurative motifs (Morrel 2000). A high society women's group and wives of high government officials took a resourceful step to open and organize camps and work centers in which *Phulkari* embroidery was used for the first time as a medium of decoration of modern household linen and furnishings (Pal 1955). The traditional art of *Phulkari* making was almost near to a lost art by the turn of the twentieth century. After the partition of Punjab during the independence movement, the domestic craft was literally lost (Brijbhushan 1990). After the partition of India and Pakistan, a development agenda encouraged automation through the machine for the locally traditional folk art of the Punjab (Pal 1955). *Phulkari* was initially done as homecraft, which was done by Punjabi women in their leisure time with a lot of love, care, and affection. This craft was not done for earning money earlier, but it was merely a result of love (Das 1992; Steel 1888). In previous *Phulkari* tradition, women used to take the silk embroidery thread and needles from village store, and in return of it, they used to give grains or hand-spun threads to the village store (Maskiell 1999). The color gives character and conveys a message, while the design tells a story and the patterns that are created bring together a harmony (Jaya Jaitly1990). According to the demand of customers, *Phulkari* embroidery is currently done from the *dupattas* into cushion covers, wall hangings, bed spreads, *kurtas*, curtains, etc. (Kaur 2011; Jaitly 2012). Man feels the need to decorate the textile through the medium of weaving, dyeing, printing, and embroidery ever since the dawn of civilization. The creation of design was followed by embroidery technique. The process of creating design is not an informal and easy way for every person. Designing is a process to use the elements with imagination, emotion, mind, and knowledge which create a meaningful statement. The word "designing" means composition of lines, forms, colors, shapes, and textures in a decorative way. This element of art is also used in textile which we used for household utility (Jyoti and Grover 2009). Textile is mostly appearing in interior spaces as cover materials for chairs, sofas, and cushions and as bed and table covers and in window management (Pile 1995). Textile sets up a feeling of softness, curvature, and flexibility into a space and also creates softness, comfort, and a feeling of being humane in hard-looking room (Pile 1995).

16.4 Role of Interior Decoration in Architecture

Architecture is the art of designing buildings and structures into a living space. It is an essential factor for the survival of human beings. It is not only essentially required for living, but it also has a remarkable value in respect of artistic representation. Historical monuments have also been considered for reckoning the status and living style of the human being. In this context, the ancient religious architecture is an easy way to trace the living style and taste of the patrons.

Art and interior decoration both are naturally related to each other because both are closely related to space construction, artistic sense, emotions, and aesthetic values. Art is related to the principles of design which make an art piece more impressive, harmonious, and balanced, and these all are necessary for interior decoration. Interior decoration has developed aspect of art. Living and nonliving spaces have always been decorated with creative and decorative techniques from ancient time to till date, which has been known as interior decoration later. Interior decoration is also very important and has a significant role in architecture. These two aspects have been related with each other from the ancient time till today. Interior designing and architecture is a part of human needs because the outer structure is particularly related to the indoor site because the outer structure is associated with the appearance of the building and indoor structures give overall support to the building. In respect of decorating and need of the building, the interior is the most important and challenging part, which should not only belong to the requirement of the individual but should also relate to the aesthetical sense. Creating a pleasurable environment through designing and planning is as old as the existence of human beings, but interior decoration is not so old. This trend has come after the mark of modern trends and is still continuing with more equipped manner.

Interior decoration has been done through various techniques and mediums. These mediums have been used on the required things in the building or house like furniture, utilities, decor items, wall decoration, other handloom items, and so on. Architectural domain particularly belongs to designing of the basic structure of a building, but the interior decoration belongs to various elements of daily needs and decoration. In this way, the outer or main structure has been designed by the architects, but the inner surface has been adjusted by the interior decorator in a balanced manner. In interior decoration, the things are not only channelized on the bases of requirement or balance, but the aesthetic approach is also created to make it more lavish and suitable. If a house is beautiful and appealing from the outside and imbalanced from the inside, the building is not considered reliable to the householder. In human life, the surroundings and ambience have also played a major role to live happily. In this way, interior decoration can be considered as the spirit of a house or office, which generated through the management of certain things in the buildings. In interior decoration, numerous things have been designed including lighting, color application, furniture placement, and other required elements of the building. Textile plays a major role in the utility items of interior furnishing. We see textile in different colors, fabrics, and shapes because textile has a variety of aesthetic beauty and noise insulation. Today, every architect wants to create new things and the environment through the use of traditional embroidered fabric in interior designing. Embroidery is the oldest technique or medium of decorating textile with an extensive history. In today's society, embroidery is not only used for ornamentation of cloths but also used to create a textural textile with the artistic expression and aesthetic. Traditional art in interior creates a royal look which attracts appreciation. *Phulkari* is also used in the interior because of commercialization before very long time back.

16.5 Interior Decoration and Textile

An architect could not create a pleasurable environment in a house without architectural design, interior decoration, and furniture. Furniture is a desirable product for people which is a necessary building block in interior spaces. People can't imagine a comfortable living style without furniture. It gives a feeling of comfort and also increases the life standard of people and helps them to maintain behavior in a comfort system (Nielson and Taylor 1994). Textile is a major element to cover the furniture, to give beautiful look, and to give relaxes during sitting or sleeping. The significance of textile has been considered through the words of Nielson and Taylor that "Fabric, more than any other tactile elements in design, has the ability to humanize our interiors. Fabric can give a sense of personal space, since so often it is selected with personal preference as a prime criterion" (Nielson and Taylor 1994). The art of decorating textiles through embroidery started very long time back. The embellishment of a fabric with threads and other material is practiced since the times of ancient Egyptians. Embroidery creates a texture which plays an important role in interior decoration. Textile designing is a part of interior decoration in contemporary trends. Textile designing has been done from the ancient time in India. Its evidences have been found from the Indus Valley civilization, and there are records of archeological survey that there was a cotton textile industry (Wilson 1979). The first linguistic records about textile designing are given in Rigveda and after that followed in numerous manuscripts. In India, textile designing has been done with various techniques on different clothes. In this popular art, fabric has been made and decorated with embroidery, dye, and other printed forms. In interior decoration, textile designing has been used on such fabrics, which have been used in furniture, curtains, and other decorative patterns. In recent times, embroidery is frequently used on the clothes for the interior decoration. Textile is a medium to cover the interior, which also includes sofa, bed, and curtain, which can give a sense of personal space and comfort. Textile is better and less expensive for interior than any other medium.

16.6 Revival of *Phulkari* in Contemporary Age

Phulkari has reached a new level with different techniques in contemporary designs and patterns. In this (Fig. 16.1) traditional forms of *Phulkari* embroidery like *Duppat* and *Shawals* have been shown with the fusion of contemporary aspect. *Phulkari* is done for earning money by Punjabi women through NGOs, government organizations, and self-organizations. With contemporary age, the art of *Phulkari* has been enhanced with various tools and techniques. *Phulkari* is now being made on various kinds of clothes unlike the traditional one. *Phulkari* embroidery is done from the front side of the cloth after printing the designs on fabrics in current scenario. Women are also using frame and other equipment to hold the cloth for fine

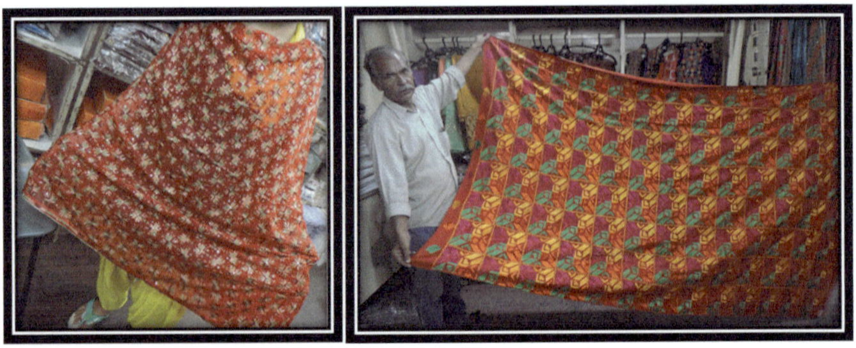

Fig. 16.1 Traditional forms of *Phulkari* (Image source: Author)

embroidery, which was not used in earlier days. In *Phulkari* art, major changes have been coming due to commercialization. Khaddar has been replaced by fine-quality fabrics like terivoiles, crepes, chiffons, lizzy bizzy cotton, stain cloth, silk cloth, etc. for the base cloth of *Phulkari*. Now *Phulkari* is available on every light and dark base color. Pat is replaced by synthetic silk thread. The thread is available in the market in every dark and light color.

As at present times, *Phulkari* designs are made according to the demand customer's choice. The design and thread color and the base color are dependent upon the choice of agent and the shopkeeper. They are perfectly varied and matched the different motifs to get an attractive design. Then the design is sent to the block and dye maker. They carved these designs with complicated details into the wooden blocks. The mango tree wood is used for the block making. The stencil blocks are sent to the tracer, who traced the design with indigo-colored ink which is known as "*Kikar ki Gond*," the mixture of ink and gum. The embossed part of the block is covered with ink and then it is traced onto the cloth. The traced sheets are sent to the embroiderer, to do embroidery with the needle and thread according to the lines of tracing patterns. Wooden frame has been used to hold the cloth for embroidery. Unlike the traditional *Phulkari*, it is not done from the reverse side of the cloth, but it is done from the front side of the cloth, and the design is also printed on the base cloth with wooden blocks. *Phulkari* is not done with the imagination and creativity. Women are using a combination of various stitches such as stem stitch, herringbone stitch, satin stitch, chain stitch, running stitch, etc. Mostly geometrical motifs are used in contemporary *Phulkari*. Flowers and birds (mostly peacock) are shown in this art of embroidery.

Phulkari is also used on the occasions of wedding by some families in Punjab for covering the bride and groom when they enter in the marriage hall. *Phulkari* is used to cover the dolly car (in which the bride comes to her in-laws' house after marriage) as decoration and to decorate the *mandap* or marriage palace.

Women are usually embroidering two to three *Phulkari*s in a month. The *Bagh* is embroidered in 10–12 days only. Today *Phulkari* is available in the market at the cost of 1,000–15,000 and *Bagh* is available at the cost of 4,000–30,000. The cost

Fig. 16.2 Products with contemporary *Phulkari* pattern (Image source: Author)

also depends upon the design and color. In this (Fig. 16.2) accessories like notepad, bottle cover, mobile cover, clutches, jutti, file cover, and handbags are also embroidered in *Phulkari* style, and the decoration of *shagun* material like *thaal* cover, *gadavi*, etc. is also being done in this style on ceremonies.

16.7 *Phulkari* and Interior Decoration

The art of *Phulkari* has remarkable features, which draws attention to art lovers because it has a harmony of aesthetical approach that attracts the viewers. In textile designing, various forms of embroidery have been used to decorate the clothes and other handloom items. Such traditional or ethnic kinds of embroideries are very popular in the contemporary Indian textile industry as well as in other countries. It has also some additional values in its application for it gives more appropriate outcomes in fewer efforts. In this way, *Phulkari* is not limited to the artistic surface, but it spreads into numerous domains of human life. The creative genius of artistic mind has been coming up with new additions and transformations of *Phulkari* in the field of urban interior decoration. The purpose behind such applications of *Phulkari* in textile designing is almost commercial and belongs to the demand of the consumers.

Fig. 16.3 Cushion cover (Image source: Author)

Phulkari has been adopted by the fashion industry as well as the interior decorators. After the development of industrialization, people have completely changed their tastes and demand of interior. They want their living space full of aesthetic beauty with simplification, which attracts every person. *Phulkari* is a form of traditional embroidery which is very colorful and attractive. *Phulkari*, which is made with vibrant colors, looks very lavish and makes people happy with an ethnic touch. The aesthetic beauty of *Phulkari* attracts interior decorator and then they use it with modern design and techniques. Forms of *Phulkari* have been used in bedsheets, cushion covers, curtains, drapes, and dress material. Moreover, the filling of interior space is also very essential to enhance its beauty. So *Phulkari* patterns have also been made on the handloom items on a large scale through handwork and machine work. In this image (Fig. 16.3), a beautiful table cover has been shown by a shopkeeper; the cover is made on a white surface with dark colors. In embroidery, white color base gave a royal and an eye-soothing look with dark embroidery because color application also gives a unique touch in textile designing. This unique touch has been given by choosing a perfect match for the interiors. Cushions are also used for comfort in the buildings or houses, which also have been made with *Phulkari* patterns (Fig. 16.4).

Such patterns are very useful for creating traditional ambience in private and commercial dwellings. In another image, a bedsheet has been displayed to show the geometrical patterns and squares in *Phulkari* style. The dark background of the sheet is perfectly matched with the colored geometrical pattern and creates a natural environment (Fig. 16.5). This kind of bedsheet seems very suitable with detailed interior space or light-colored walls. In recent trend sofa cover is also embroidered with the geometrical patterns and vibrant colors. In traditional Punjab *Phulkari*-embroidered sofa cover was used by women for a special guest. In this (Fig. 16.6) sofa/chair cover has been shown in various brightly colored patterns on light-colored base cloth which creates balance in dark interior space.

Fig. 16.4 Dining table covers (Image source: Author)

Fig. 16.5 Bedsheet (Image source: Author)

Curtains are also made with flower motifs of *Phulkari*, which give an effect of ethnicity with graceful appearance (Fig. 16.7). From the one hand, it looks very ethnic and covers the space and prevents buildings from dust and sunlight on the other. It can be a very beautiful medium to decorate the walls of the interiors through geometrical patterns after some experiments. The use of *Phulkari* on the wall looks like wall art. In recent trends wall art is very popular. It has left a good impression of beautifying the space through vibrant colors and geometrical motifs to the onlookers. The decoration of *Phulkari*-embroidered cloth on the wall (Fig. 16.8) creates royal and ethnic look.

In the next image (Fig. 16.9), an image of a room's interior has been shown to highlight the placement of embroidered designs. The lamp, which is placed in a corner, is looking very dull and empty, but if the blankness of the lamp is being

Fig. 16.6 Sofa/chair cover (Image source: Author)

Fig. 16.7 *Phulkari* on wall (Image source: https://www.pinterest.com/lenacrosell/textiliertextiles/. http://indigochreblog.com/2012/12/18/before-and-after-a-prospect-heights-brownstones-resurgence-part-four/dsc_2537-112/)

replaced with *Phulkari* designs, it will look very appealing. So, the lampshade in the corner of the image can be considered to evaluate the beauty of *Phulkari* designs. In another image, the blankness of the table can also be filled with the lamp, which is also shown in the corner of the image. In this way, numerous items can be ornamented with *Phulkari* as per the choice of individuals either to utilize the space or to make it beautiful.

Fig. 16.8 *Phulkari* on curtain (Image source: https://www.pinterest.com/lenacrosell/textiliertextiles/. http://indigochreblog.com/2012/12/18/before-and-after-a-prospect-heights-brownstones-resurgence-part-four/dsc_2537-112/)

Fig. 16.9 *Phulkari*-embroidered lamp (Image source: http://medesignwe.com/content/embroidered-lamps-modern-interiors)

16.8 Some Experiments with *Phulkari* Patterns

The revival of *Phulkari* can be seen in the previous image, but it can also be modified in other things of interior decoration. In an image (Fig. 16.10), an experiment of *Phulkari* design has been done on the sofa and the other sofa has been left blank to show the difference. The sofa with *Phulkari* patterns has been looking more gracefully ethnic than the other one. *Phulkari* patterns on the sofa set have given a colorful and bright environment and change the whole room's look. It has given pleasure to the eyes of onlookers. It has given royal atmosphere.

In another experiment (Fig. 16.11), some images of the lamp and *Kandil* have been taken into consideration to show the harmony of geometrical patterns, which

Fig. 16.10 *Phulkari* designs on sofa (Image source: Author)

Fig. 16.11 Lampshades with *Phulkari* patterns (Image source: Author)

can be used to fill the blankness of the space or to cope up with the dull and bright color scheme. These lamps may be made in geometrical patterns as wall hangings. These images with *Phulkari* patterns give a rich environment to the whole living space. Lamp with *Phulkari* pattern creates a nice mood and given a luxurious application into the furniture.

In this image given below (Fig. 16.12), another experiment has been done on the wall hanging which is used to decorate the room and also used for welcoming the guest in the house. On the upper part of the wall hanging, the *Phulkari* patterns of the geometrical flower pattern has given exquisite look to the living space. The beautiful, bright, and vivid colors fill the blank space on the door.

Fig. 16.12 Wall hanging with *Phulkari* patterns (Image source: Author)

16.9 Conclusion

To conclude *Phulkari* has been done in spare time by the women of Punjab in earlier times, but now it is used to empower them. *Phulkari* textile can come up with the new dimension of commercial values if its rejuvenation will be done in interior decoration, for instance, it can apply on the ceilings, walls, handloom items, crockery, and so on. In this way, marginalized women may be more involved in this art rather than other laborious work and get an exposure in interior decoration directly or indirectly. Furthermore, *Phulkari* application in textile designing has also come up with various dimensions. In interior decoration, it satisfies the traditional thrust of the consumer because in Indian tradition, people have been connected with their cultural roots directly or indirectly. Such artistic revival gives them an eye-soothing and relaxing ambience and a chance to show their ethnicity. On the other hand, such interior decoration also provides a natural environment for it has vegetation and floral motifs. People can feel more connected to the nature through these decorative motifs of *Phulkari*. It has been developed very much and reached in almost all domains of decoration including clothes, accessories, and home decor, but it can also be used on other things like lamps, on walls, on furniture, and so on. In this context the edited images look more graceful and ethnic than the blank one. This wary implementation of *Phulkari* pattern may vary according to the demand of the consumers or it can be dependent upon the necessity space, making, for instance, an individual wants to renew his or her less decorative place and *Phulkari* decorative and colorful motif could be a better medium. The patterns of *Phulkari* can also be drawn into the interior of the roofs in the form of boundaries, circles, and other geometrical designs, which may be a better way to give the interior an ethnic look with colorful embellishment. It can also be applied in the form of murals to decorate the surface as per the choice of the consumer. Thus, *Phulkari* could be an excellent ethnic embellishment in interior decoration. *Phulkari* patterns on modern interior are fresh and new idea to give royal, luxurious, bright, colorful look to the living space.

References

Aryan K (1983) The cultural heritage of Punjab: 3000 B.C. to 1947 A.D. Rekha Prakashan, Delhi

Banerji A (1955) *Phulkaris*: a folk art of Punjab. Marg 8(3):58–65

Barnard JG (1991) Traditional Indian textiles. Thames & Hudson Ltd., London

Brijbhushan J (1990) Indian embroidery. Director, Publication Division Ministry of Information and Broadcasting Government of India, Delhi

Chattopadhyay K (1975) Handicrafts of India. Indian Council for Cultural, New Delhi

Crill R (1999) Indian embroidery. Indian Embroidery, London

Das S (1992) Fabric art: heritage of India. Abhinav Publications, New Delhi

Dhamija J (2004) Embroidery: an expression of women's creativity. In: Dhamija J (ed) In Asian embroidery. Abhinav Publications, Delhi, pp 17–24

Dhamija J (2007) Embroidered gardens of flowers: Bagh and *Phulkari* of Punjab. Marg 58(4):13–21

Dongerkery K (1951) The romance of Indian embroidery. Thacker and company Ltd., Bombay

Dutta DM (1985) A catalogue on Phulkari textiles in the collection of the Indian Museum. The Museum, Calcutta

Gillow J, Barnard N (1991) Traditional Indian textiles. Thames & Hudson Ltd., London

Harvey J (2002) Traditional textiles of Central Asia. Thames and Hudson, London

Himadri B (1982) Agrarian society of the Punjab: 1849–1901. Manohar, New Delhi

Hitkari SS (1980) Designs and patterns in Phulkaris. Phulkari Publications, New Delhi

Jaitly J (2012) Craft atlas of India. Niyogi Books, New Delhi

Jaitly J (1990) The craft traditions of India. Lustre Press, New Delhi

Jyoti, Grover E (2009) Designing and printing of bed cover using CAD technology. Text Trends 59(7):29–32

Kaur P (2011) Revival of Punjab's traditional handicraft: *Phulkari*. Asian J Manag 2(1):28–38

Lehri R (Super Book House) (2006) Indian embroidery- ethnic & beyond India. Super Book House, Mumbai

Maskiell M (1999) Embroidering the past: *Phulkari* textiles and gendered work as "tradition" and "heritage" in colonial and contemporary Punjab. J Asian Stud 58(2):361–388

Morrel A (U.S.) (2000) The techniques of Indian embroidery. Diane Publishing Co., Collingdale

Naik SD (1996) Embroidery and traditional handloom weavers. A.P.H. Publishing Corporation, New Delhi

Nielson KJ, Taylor DA (1994) Interiors an introduction. McGraw-Hill, Boston

Paine M (1990) Textile classics. Mitchell Beazley Publishers, London

Pal R (1955) The Phulkari-a lost craft. National Printing Works, New Delhi

Pile J (1995) Interior design, 2nd edn. Harry N. Abrams, Inc., NewYork

Steel FA (1888) *Phulkari* work in the Punjab. J Indian Art Ind 2:71–72

Tondon P (1961) Punjabi century, 1857–1947. Hind Pocket Books (P) Ltd, Delhi

Townsend WG (1907) Embroidery or the craft of the needle. Trustlove, London

Wilson K (1979) A history of textile. The University of Michigan: Westview Press, Boulder

Chapter 17
Religion Interacts with New Urbanism Holistic City Anandpur Sahib

Vikas Chand Sharma and Sunita Vimal

Abstract Both ancient and modern cities are the result of a limited range of configurations that structure human action in concentrated populations. Anandpur city is a religious city, which represents attributes beyond its natural characteristics. A corollary of the development of communities in Anandpur Sahib has been a steady growth in the quantity and aesthetic presence of Gurdwaras within Sikh urban landscapes. Strategies to develop these buildings have frequently given rise to forms of aesthetic constitution that are embedded in processes of identity construction among new cities. Taking the religious impact and emerging urbanism as case studies, this paper demonstrates the extent to which urban planning processes condense and mediates the relations among social groups. In addition, it explores the changing emphasis of the city relating to adore places, as these have shifted from restriction to multicultural "celebration."

Elements convening definitions of the city Anandpur Sahib are:

Heterogeneous societies and discrete buildings as economic and administrative, social, institutional neighborhoods and associated personnel
Compacted and overlapped packing of residential and nonresidential structures
Monumental core of unique buildings (e.g., Keshgarh Sahib Gurdwara, bus stand structure)
Five forts of the city
The Khalsa Heritage Museum
Special characteristic feature "City profile" of Anandpur Sahib that shows maximum building height at the center of the city and less height as one moves away from the city center, central focuses the enshrined center, whose access was restricted and where Gurdwaras predominated

---

Public Places, Urban Spaces: The Dimensions of Urban Design by Matthew Carmona (pages 40–42)

V.C. Sharma (✉)
School of Architecture, MM University, Sadopur, Ambala, Haryana, India
e-mail: ar.vikassharma@gmail.com

S. Vimal
COA, Bhaddal Technical Campus, Ropar, Punjab, India
e-mail: svimal30@gmail.com

F. Seta et al. (eds.), *Understanding Built Environment*, Springer Transactions in Civil and Environmental Engineering, DOI 10.1007/978-981-10-2138-1_17

Keywords New urbanism • City fabric • Advocacy planning • Religion and city growth

17.1 Introduction

Anandpur Sahib, "City of Bliss,"[1] is one of the most holy places of the Sikhs, maintaining and preserving their religious traditions, customs, and history. It has a number of historical Gurdwaras (Fig. 17.3), and Nangal and Bhakra projects are located at 20 km to the north and mark the importance of the city. These projects have helped to glorify the significance of Anandpur.

17.1.1 Formation of the City (Historical Attributes)

Why did cities form in the beginning? Little evidence asserts the conditions which gave rise to the birth of cities, whereas some theorists have hypothesized on what they consider pre-estate and basic structure that could explain the rise of cities. At the beginning, the Anandpur zone, from Kiratpur Sahib to Nangal, was a crammed forest, surrounded by a lower belt of Shivalik hills, river Satluj, Charan Ganga, and

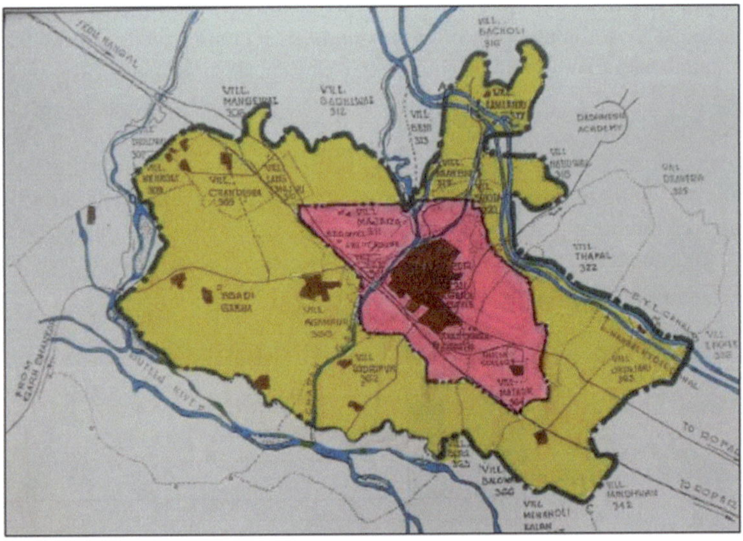

Fig. 17.1 Physical map of Anandpur Sahib showing the road network and other settlements nearby (geographical map)

[1] http://www.nalandainternational.org/index.php/anandpur-sahib (Accessed on 08/10/2015).

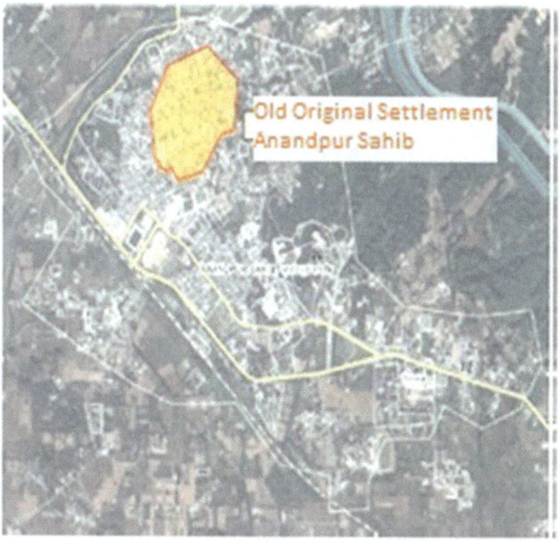

Fig. 17.2 Original settlement in Anandpur Sahib (1925)

other rivulets (Fig. 17.1). This parcel of land is about 45 km in length and 10–12 km in width without any human population (Fig. 17.2).

At the beginning, the settlement of the city was around five,[2,3] Gurdwaras as mentioned in Fig. 17.3. People had chosen to build their dwellings within a close proximity to these Gurdwaras. Population factor which directs continuous growth and city fabric of this settlement is identified by various recordings. During 1800–1810, the population of Anandpur Sahib was approximately 2800, whereas the first census recorded it 6869 in 1868. Further, during an epidemic attack in the twentieth century, it was recorded around 7000[4] (Fig. 17.2). At this time an epidemic spread in the town and the adjoining villages, and migration of city root dwellers took place (Fig. 17.4).

17.2 New Urbanism

The execution of new projects at Anandpur Sahib is likely to give a new look to the town. The arrival of global tourists and pilgrims is going to affect the development both in terms of planning and infrastructure. The city of the period of Guru Sahib has entered into conflict of two major schools of thoughts conservation and new developments. This will probably add new dimensions toward developments. Annual Sikh gathering which took place during the fest "Holla Mohalla"[5] attracts around two lakh visitors. The city starts to find self-identity and potential to deal with,

[2] *The Sikhs* By Gene R. Thursb (pages 17–19).

[3] http://www.sikhiwiki.org/index.php/Anandpur [Accessed 13 February 2015].

[4] Office of the registrar general and census (web). Available at censusindia.gov.in/ [Accessed 13 February 2015].

[5] *Civil Society in Malerkotla, Punjab: Fostering Resilience Through Religion by* Karenjot Bhangoo Randhawa.

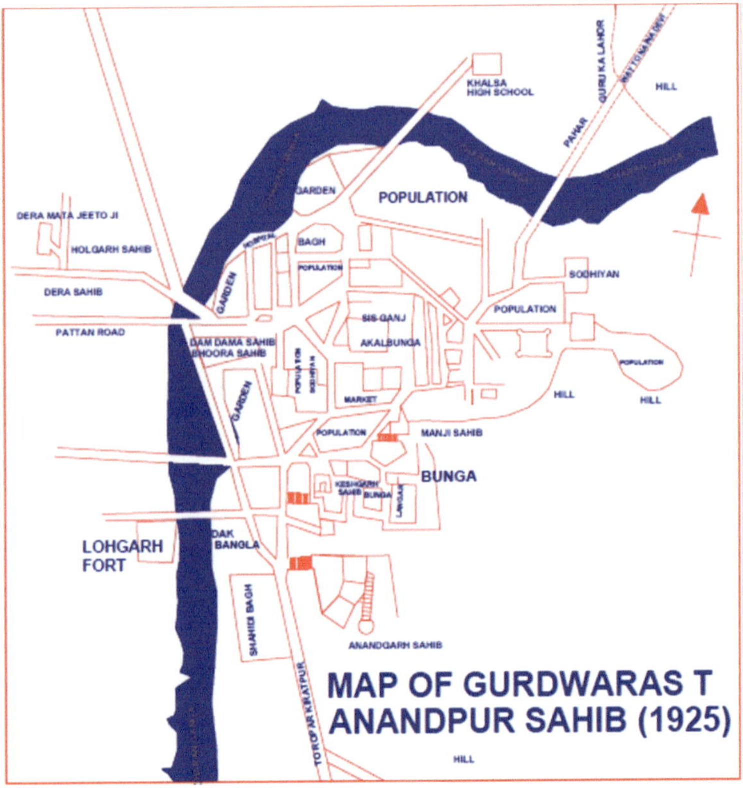

Fig. 17.3 Evolution of the city (Source: saty.co.uk/Sikh-history/anandpur_sahib.htm)

Fig. 17.4 Google earth image of Anandpur Sahib

Fig. 17.5 Current settlement in Anandpur Sahib (2014)

whereas Anandpur Sahib stands for its originality which already engaged with religion. A tourist is not a single element who stays and goes back, but he brings a wide scope to business to both local crafts and essential amenities.

Accessed elements of New Urbanism are identified and being refreshed day by day:

Environmental sustainability[6] promoted by ecological characteristics of semi-hilly terrain, conservation in terms of Sikh heritage forts and correlated characteristics

An intervention of technology as being narrated by PHPTB[7] and government organizations, a complex facade of various infrastructures (commercial, residential, institutional, etc.) which is capable for desired efficiency

Preservation of their customs and festivals with public participations

An efficient execution of same provoking social access and transit-oriented development too (Fig. 17.5)

Apart from regional integration that existed, human scale and institution integrity are defining a New Urbanism. The awareness of integrity and how the intervention of religion as an element in the overall web formation affect all major attractions to choose the city as an example of New Urbanism.

---

[6] *Sustainable Urban Design: An Environmental Approach edited* by Adam Ritchie, Randall Thomas.

[7] Punjab Heritage and Tourism Promotion Board.

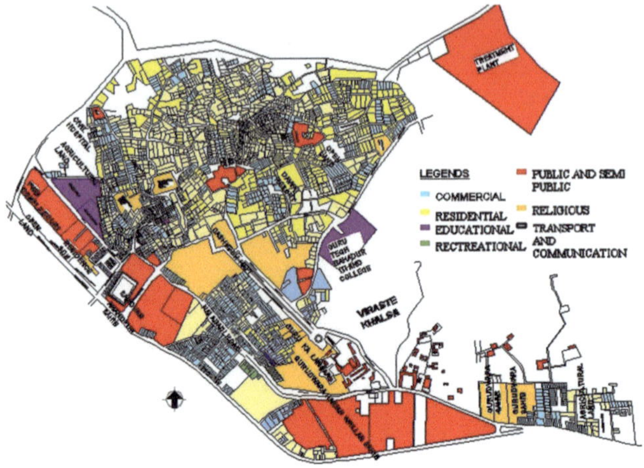

Fig. 17.6 Land-use plan Anandpur Sahib

17.3 Advocacy Planning[8]

It is a peaceful zone consisting prosperity as plenty of fertile land (Fig. 17.4) capable to yield two crops annually. This allows Anandpur Sahib to hold a rank among the cities of influential GDP (per capita \$3900). Hence, the changing city is capable to become self-sufficient in terms of employment and sources require for the same. Dominating forces to the city image of Anandpur Sahib of the eighteenth century considered are a number of new buildings in and around the city. However, basic demographic of the city is matching to the developments and almost all the shrine zone is preserved.

Anandpur Sahib provokes the models:

(i) It focuses on residential areas in which people of similar origin and occupation as an agriculturist tend to live in the same neighborhood yet wanted to live with others of similar origin or income; various parts of the city came to be identified with a particular social class or ethnic group.
(ii) There are differences in land use (Fig. 17.6) near the center of a city around the main shrine; these differences would often be maintained as the city expanded outward (Hoyt's model[9]).

[8] Advocacy planning is another radical departure from past theoretical models. This model takes the perspective that there are large inequalities in the political system and in the bargaining process between groups that result in large numbers of people unorganized and unrepresented in the process. It concerns itself with ensuring that all people are equally represented in the planning process by advocating for the interests of the underprivileged and seeking social change (Mazziotti 1982).

[9] The *sector model*, also known as the *Hoyt model*, is a model of urban land use proposed in 1939 by economist Homer Hoyt. It is a modification of the concentric zone model of city development (Hoyt 1939).

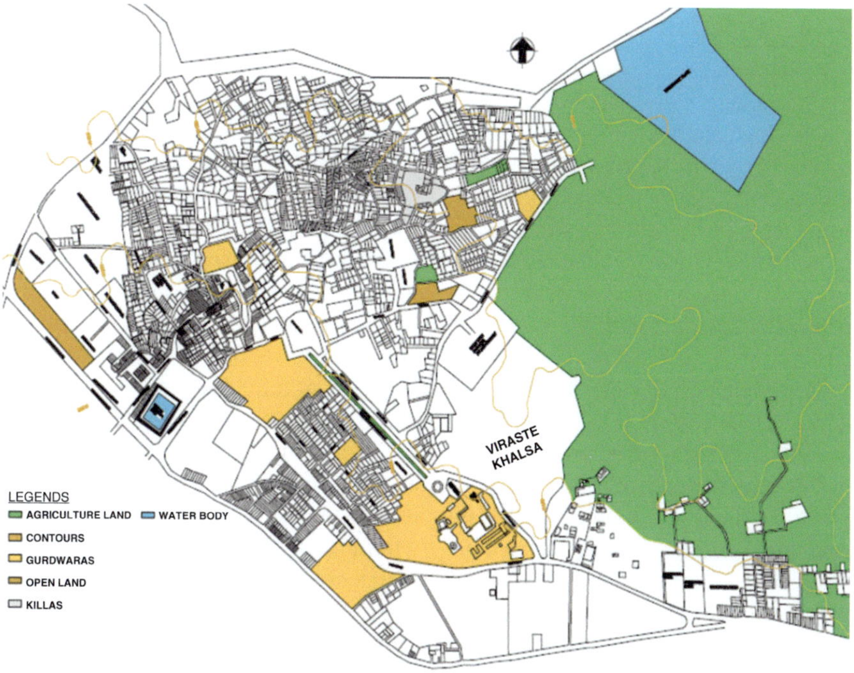

Fig. 17.7 Base map Anandpur Sahib

(iii) In an initial stage, APS model poses inequalities in the political system, and land distribution methodology has been adopted in the bargaining process among groups. This may be observed as unorganized and unrepresented distribution of a number of people, which remains the major outcome (Fig. 17.7).

Concern to ensure advocacy for people, their interests, and probability of changes in social ground are equally reflected in the new planning development. This is responsible for the future growth in accordance with the existing scenarios for infrastructure and built environment. A visible vibe of urbanity is nourished by the incoming of global tourists, interaction with new dimensions, and promotion new defined sector modules. Anandpur Sahib defines the assumed plurality of public interests in which unrepresented groups also became part of the planning and overall public participations on conceptual stage remain its driven forces. Advocacy planning at Anandpur Sahib may be defined as a radical departure from past theoretical models.

17.4 Religion and City Growth

The term Holy City[10] applies to both types of cities, one central to the history or cities which has the dominating attempt to motivate additional growth and the other related to specific religions. History governs that three forces "the sacred, the secure, and the busy" work together to form a city.[11] The earliest cities from every culture were centered on a temple and reinforced by a shared religious belief. An example of Madurai city can be cited here, which has been inhabited since the third century BCE. After the emergence of the intervention of imperial powers and the growing population, the sacred space was shifted from various smaller towns to a centralized city.

Anandpur Sahib contains a headquarter complex of religious edifices (Figs. 17.8 and 17.9) which acts as a pivot to the city growth. To protect and regulate this, urban development authorities have restricted new constructions with the close proximity of this center from two prescribed points of views, supplementary character of religious buildings and a transparent zoning of the city. The central location of these edifices adversely promotes the incoming of traffic (Pilgrims and visitors) or a major destination of human traffic. This touches the intent of this paper at which level religion and its existence affect the city growth. An approach to center defines a layout pattern of the following streets, arteries and major roads, and standby landscape.

This festival "Holla Mohalla" commences the legacy of Sikhism and Khalsa.[12] Folk arts and cultural events take place along with native game of state Kabaddi. There is a myth that on this occasion, you can see the power and blessings (pertaining to growth) all around. A large quantity of pilgrims around the two lakhs participates in this festival from all over India and abroad (Fig. 17.8). This type of gathering can quest to the city planning and its efficiency. Anandpur Sahib has populations like Sikh, Hindu, and migrates from Pakistan preferably Muslim communities. Somehow, their individual togetherness reinforces social division and the sense of separation, which is visible in organic-like lattice (Fig. 17.7).

On commerce ground they tend to intact with each other the result module like commercial spaces (Fig. 17.6) equally reinforced by adequate urban elements of the city image. A need of New Urbanism can be felt here just to connect ethnic and religious webs. Frequency of adaptation must be controlled by their local customs and sensitiveness at all. Sensitiveness here recalls the magnitude of large span

---

[10] *Holy city* is a term applied to many cities, all of them central to the history or faith of specific religions. Such cities may also contain at least one headquarter complex (often containing a religious edifice, seminary, shrine, residence of the leading cleric of the religion, and/or chambers of the religious leadership's offices) which constitutes a major destination of human traffic, or pilgrimage to the city, especially for major ceremonies and observances (https://en.wikipedia.org/wiki/Holy_city).

[11] Kotkin (2006).

[12] The *Khalsa* is the collective body of all initiated Sikhs represented by the five beloved ones and can be called the Guru Panth, the embodiment of the Guru (Wikipedia.org).

Fig. 17.8 Ritual activities in Anandpur Sahib

Fig. 17.9 Gurdwara Shri Keshgarh Sahib (Source:saty.co.uk/sikhhistory)

projects and their enforcement in the city planning. Khalsa Museum[13] (Fig. 17.10) designed by Ar. Moshe Safdie brings example, as how it serves to attract tourists and pilgrims. This results a consultation between religion and the emerging need in building environment. On one side, it promotes handcrafts to locals as well as nurturing sense of heritage, other side it recalls to infinity by the volumetric interference of existing skyline is another phase of a visible urbanism dilemma.

Museum campus is responding to need of the population, providing business to the locals and makes the city marked on urban literature globally. PHTPB paid to

[13] http://www.msafdie.com/#/projects/khalsaheritagecentre (Accessed on 08/03/2015).

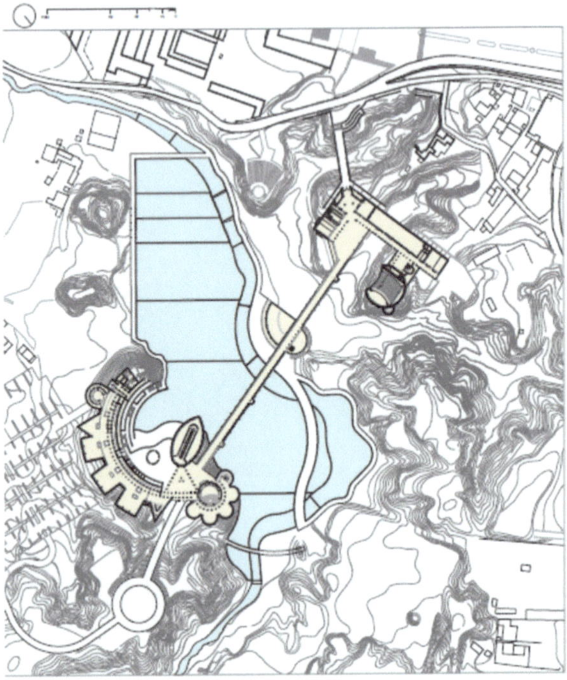

Fig. 17.10 Khalsa heritage museum at Anandpur Sahib (site plan) (Sources: Moshe Safdie Associates)

Fig. 17.11 Multipurpose space (Gurdwara Committee Land)

have it installed in order to attract worldwide tourism. Open spaces which are going to be used by ritual activities during ceremonies and festivals also serve as alternative parking grounds, reserved grounds for political rallies (Fig. 17.11) which bring the intact ingredient of Indian cities together, "The Interaction" (Fig. 17.12).

Fig. 17.12 Interaction in the public realm (Parking at Khalsa Heritage Museum)

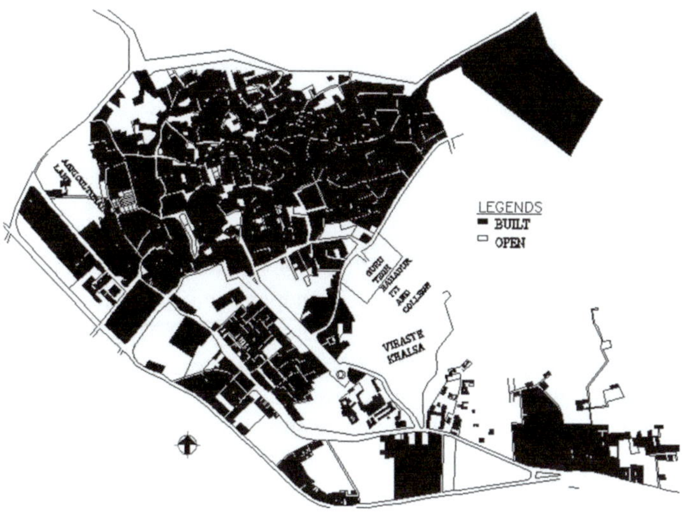

Fig. 17.13 Built-open plan Anandpur Sahib

17.5 Findings at a Glance

- The traditional urban web was preserved (Fig. 17.13) in the whole area of the old city, and any interventions (widening, alignments, or arcades) in its planning are subjected to the laws of the region.
- Merging of land parcels seems to be flexible, and their existing size is subject to the change based on the sizes of the buildings and consequently the town image.
- Emerging needs of new infrastructure are guided by the existing fabric identity of the city. A newly built bus stand located on an angular axis within the center

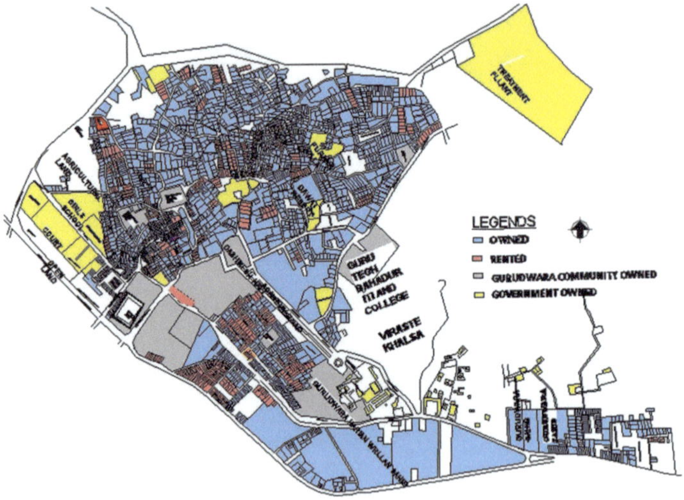

Fig. 17.14 Property map Anandpur Sahib

of urban lattice is in corelation to the façade treatment of central edifices and is a live example of preservation-based growth. This scenario again provokes the statement, "Any planning, innovation in a city must be governed by the logic of the whole and in design terms must offer a formal response to pre – existing spatial conditions" (Rob Krier).[14]

- Apart from the material dimension of the city, the special objective of the present paper is the immaterial dimension of Anandpur Sahib, comprising its socioeconomic content and its cultural features, which enables the main ingredient of the city's image and uniqueness.
- Apart from the global history of the city formation, documentation shows that Anandpur Sahib was a sparsely populated city and a flexible trade center and still maintained originality or primarily religious focus (Fig. 17.14).

17.6 Conclusion

Today, cities are facing congestion due to heavy migration in urban areas. To retain identity of the city and legacy of vision on which the city has been formed, it became necessary to find the fractional module of the particular city. Due to the change in lifestyles and living patterns, the overall structure of the cities whether organic or inorganic is going to be affected day by day. It needs to be readdressed or redefined

[14] *Ecopolis: Conceptualizing and Defining Sustainable Design edited* by Dimitra Babalis (page 86).

in corelation with new emerging theories of urban design. Vitality must be encompassed along with the focus on similarities in pattern and function.

We therefore decided to document the city of bliss Anandpur Sahib in relation to related elements of urban design. This paper explores the historical and ongoing influence of religion on urban planning, space utilization, urban identities, and communities. It argues that the conceptual and empirical approaches to planning cities in India need to be developed out of analytical concepts that preserve the local sense and identity of the place.[15] Issues of urbanism like traffic and transportation, implementation of landscaping design and building design theories, environmental impact, viability of the street, techniques for new development, and connection between new urbanism and regional planning are nurturing the city Anandpur Sahib and making the vitality manifold against the unwelcome development which leads to a decline in interaction and also in architecture.

The quality and variety of architecture in existing cities can be enhanced and protected by the implication of New Urbanism and related theories. As a solution of resettlement and migration issues, urban design is being utilized by many cities. The study of Anandpur Sahib shows how profoundly variables (social, economic, and cultural) are interconnecting with the built environment. It's like a recall to the connection of urban design and communities.

References

Arshi P (1986) Sikh architecture in Punjab. Intellectual Pub. House, New Delhi

Babalis D (2007) Ecopolis: conceptualising and defining sustainable design. Alinea international, Firenze, p 86

Carmona M (2003) Public places, urban spaces: the dimensions of urban design. Architectural Press, London

Daljeet (2004) The Sikh heritage (a search for totality). Prakash Pub., New Delhi

Database of Geographic Coordinate Information. Available at Tageo.com. Accessed 15 Mar 2015

Delimitation Commission of India (web). Available at eci.nic.in/delim/. Accessed 13 Feb 2015

Hoyt H (1939) The structure and growth of residential neighbourhoods in American cities Washington. Federal Housing Administration, Washington, DC

Kotkin J (2006) THE CITY: a global history- modern library. Modern Library Chronicles, London

Krier R (2006) Town spaces. 2nd edn. Birkhäuser Architecture, Boston.

Low SM (1996) The anthropology of cities: imagining and theorizing the city. Annu Rev Anthropol 25:383–409

Mazziotti DF (1982) The underlying assumptions of advocacy planning: pluralism and reform. In: Paris C (ed) Critical readings in planning theory. Pergamon Press, New York, pp 207–227

Narayanan Y (2014) Religion, heritage and the sustainable city: Hinduism and urbanisation in Jaipur (Routledge Research in Religion and Development). Routledge, Oxon

Office of the Registrar General and Census (web). Available at censusindia.gov.in/. Accessed 13 Feb 2015

[15] *Religion, Heritage, and the Sustainable City*: Hinduism and Urbanisation in Jaipur by Yamini Narayanan.

Randhawa K (2012) Civil society in Malerkotla, Punjab: fostering resilience through religion. Lexington Books. Lanham, MD
Singh H (2015) Structures in architecture. Abhishek Publication, Chandigarh
Thomas R, Ritchie A (2008) Sustainable urban design: an environmental approach. Taylor & Francis, Oxford
www.intercultural.dk (Internet Source)
www.theatlanticcities.com (Internet Source)

Chapter 18
A Study on Geometrical Motifs with Special Reference to Old *Havelis* of Saharanpur

Aayushi Verma and Ila Gupta

Abstract Art and architecture demonstrate fascinating phenomenon, created in various geometrical forms and obtained in both architecture and craft. These geometrical forms have been traced from the prehistoric cave paintings and refined during the Islamic and the Mughal reigns. Old *havelis* of Saharanpur, over a period of time, display the glimpse of Islamic architecture with architecture treatment and structural ornamentation. The term *havelis* is generally used for private and huge mansion where people live together. In this context, the tremendous variety of geometrical motifs has been introduced in old *havelis* of Saharanpur which can be seen on the structural part of old *havelis* like doors, brackets, windows, and lattice screens in two-dimensional space with Islamic beliefs and skillful manner of artisans. In fact, variation is the most distinctive characteristic of the geometrical design which provides a high-level experience of aesthetics which balances simplicity and complexity that creates harmonious ambience in human life. These wide varieties of decorations demonstrate the glimpse of Indian art as well as the mysteries of Islamic art. Such geometrical designs generate spiritual and lyrical ambience and mesmerize the viewers through marvelous variety of patterns. This paper attempts to examine the geometrical ornamentation in old *havelis* of Saharanpur with special reference to the symbolic value of geometry in human life. Additionally, it briefly investigates and identifies the rationale behind the application of design principles in the intricate geometrical patterns. It also elaborates how these patterns enhance the beauty and attraction of old *havelis* of Saharanpur.

Keywords Geometrical motifs • *Havelis* • Aesthetic • Saharanpur • Ambience

A. Verma (✉)
Department of Humanities and Social Sciences, IIT Roorkee, Roorkee, India
e-mail: ayushi03verma@gmail.com; ilafap@gmail.com

I. Gupta
Department of Architecture & Planning, Indian Institute of Technology Roorkee, Roorkee 247667, Uttarakhand, India
e-mail: ilafap@gmail.com

F. Seta et al. (eds.), *Understanding Built Environment*, Springer Transactions in Civil and Environmental Engineering, DOI 10.1007/978-981-10-2138-1_18

18.1 Introduction

Art and architecture are considered as the composition of creativity and imagination which required technical skill and aesthetic consideration. Architecture denotes the basic need of shelter but represents the time, era, culture, and communities in which decorative elements are lavishly adopted for beautification and embellishment of buildings. Decoration is an enchanting and memorizing phenomenon of art which is strongly associated with architecture through wall decoration, mural painting, wood or stone carving, and so on. In this context, the decorative elements are prominently dominated by narrative components which reveal a story behind the application and enrich the grace of architecture. In architecture, design elements have ability to play a key role through their schematic image, where spectators achieve self-gratification and immense pleasure. In the current scenario, exterior and interior decorations play a very significant role in society which should be very attractive and impressive. However exterior ornamentation creates a welcoming effect on the observers. Exterior and interior decorations are considered as kinds of aesthetic movement, in which aesthetic pleasure comes through the decoration of buildings. In this context, Saharanpur has an illustrious history and a glorious past of wood carving. With the passage of time, the historic city of Saharanpur got attention for the old *havelis*, in which geometrical motifs were lavishly adopted for the surface decoration as well as filling the surface with stone and wood carvings in interior and exterior decorations. These *havelis* demonstrate a vast vocabulary of design elements on architectural surface which enhance aesthetic notion and create lyrical compatibility in society.

18.2 Research Methodology

The research study is conducted on geometrical motifs which have been applied in the exterior and interior of old *havelis* of Saharanpur. In this regard, five old *havelis* have been selected for findings: *haveli* in *chhatta* Jambudas, *haveli* in *chhatta* Barumal, *haveli* of Aatma Ram, *haveli* of Nanda Ghati, and *haveli* near *chhatta* Jambudas. The investigation consists of a comprehensive review of relevant literature, definitions' selection of the case study, and field works.

18.3 Study Site

The Ganga-Yamuna Doab region in western Uttar Pradesh, where Saharanpur is situated, is known for artistic creativity and aesthetic pleasure. In this region, Saharanpur is one of the prominent centers of wood carving on national and international level. Saharanpur got its name from the name of Sufi saint "Shah Harun Chisti" in the thirteenth century. It was ruled by the Maurya dynasty, Gupta dynasty, Gujjar, Pratihar, Sultanate, Lodhi, Mughal, and Maratha and, respectively, came under British rulers. During the Mughal Period, Akbar was the most noteworthy ruler who made Saharanpur an administrative unit under the province of Delhi and appointed a *jagir* of the tract who was Raja Shah Ranbir Singh. He was the Mughal treasurer and belonged to the Agrawal Jain community and also the known founder of Saharanpur. He was the prominent governor of Saharanpur, and during his tenure, some craftsmen from different places were attracted to Saharanpur due to peaceful market and finally settled here (Hāṇḍā and Jain 2000). Some craftsmen migrated to Saharanpur from Kashmir as early as 400 years ago (SWCA). In Saharanpur, most of the renowned *havelis* were built by the wealthy Jain community, in which Hindu and Islamic design elements were applied as the ornamentations of exterior and interior of buildings. These *havelis*, enriched with wall paintings as well as wood and stone carvings, became a part of cultural heritage and an irreplaceable source of life and inspiration. The old *havelis* of Saharanpur were considered as status symbols of wealthy Jain families. With the passage of time, some *havelis* have been demolished, and some of *havelis* are in dilapidated condition. Three *havelis* are situated in *chhatta* Jambudas and *chhatta* Barumal. The other two *havelis* are situated in the main market such as the *haveli* of Aatma Ram and Nanda Ghati (Figs. 18.1, 18.2, 18.3, 18.4, and 18.5).

Fig. 18.1 *Haveli* in *chhatta* Barumal

Fig. 18.2 *Haveli* in *chhatta* Jambudas

Fig. 18.3 *Haveli* of Aatma Ram

Fig. 18.4 *Haveli* of Nanda Ghati

Fig. 18.5 *Haveli* in *chhatta* Jambudas (Source: Author)

Fig. 18.6 A general view of *chhatta* Jambudas (Source: Author)

Chhatta literally means the extension of *chhat* (roof). It is covered with a wooden overbridge that connects two *havelis*, and such bridges run across the streets without hindering the path below. The *chhattas* are the hallmark of Saharanpur streets. With the passage of time, *chhatta* Jambudas has already lost its originality, and due to lack of conservation, it has been reconstructed (Fig. 18.6). Nowadays, *chhatta* Jambudas *haveli* is used for marriage functions by needy families who have very low income and whose budget does not allow to hire a royal marriage banquet hall (Fig. 18.2). Therefore, its façade is very beautiful although maintenance is required. Due to commercialization, most of the old *havelis* have been demolished and transformed into the markets. For instance, the *haveli* of Nanda Ghati is in very poor condition and has been converted into a commercial area where several shops have been constructed for keeping the commercial view in mind (Fig. 18.4). Some parts of that *haveli* still remain which are neglected (Figs. 18.6 and 18.7).

18.4 Design Elements of Old *Havelis*

Old *havelis* of Saharanpur reveal a vast vocabulary of design elements such as auspicious motifs, figurative idols, flora and fauna and a rich repertoire of geometrical motifs through carving. Over a period of time, these motifs have developed their own identity and position in the society through their visual imagery and visual communication language and give aesthetic pleasure through their visual forms. Currently, old *havelis* of Saharanpur are acknowledged for their prominent and meaningful motifs which manifest their motives and aspiration through symbolic image and underlying meanings. In this regard, geometrical motifs capture a vital place. It is not just a basic shape of geometry but holds its own identity in Indian as

Fig. 18.7 A general view of *chhatta* Barumal

Fig. 18.8 Square grid pattern

well as Islamic art. This form of art is strongly inspired by mathematics and science. Old *havelis* of Saharanpur convey the keen knowledge through the geometrical motifs and interpret ideologies of Indian and Islamic beliefs. In old *havelis*, the geometrical designs demonstrate the different forms of geometrical calculation such as the square grid, compound patterns, isometric grid, hexagon, star, octagon, and overlay and open lattice patterns (Fig. 18.8 and 18.9). The development of geometrical patterns reached a degree of complexity and sophistication which is the marvelous and harmonious attribute of old *havelis*. The geometrical patterns can be seen on wooden doors, door frames, brackets, windows, railings, and all exteriors and exterior surfaces of old *havelis*.

Fig. 18.9 Hexagon pattern

18.5 Geometry

Geometry is a combination of mathematics and science, recognized as a powerful organizational tool to describe the visual study of shapes, sizes, patterns, and positions. Since ages, geometry has been used for aesthetical, functional, and engineering purposes, which creates a layout of overall composition. In art, geometry is an essential equipment which constructs the basic form of design. Geometry and art have a long relationship judging from the very first evidence of excavation of prehistoric civilization. Geometry over the passage of time has evolved during different eras owing to the aesthetic sensibility of different monarchs. Refinement in these geometrical patterns is to be seen under the patronage of different Islamic rulers. One of the major influences was that of the Mughals.

18.5.1 Application of Geometry in Indian Art and Architecture

Throughout the world, Hindu architecture is considered as a glory of India which throws the light on *Vastu-Shastra*. *Vastu-Shastra* is known as a science of architecture and construction which deals with design, layout, space arrangements as well as spatial geometry, and so on, and it is based on *Vastu-Mandala*. "*Vastu-Mandal*a is recognized as a geometric representation of the cosmos in forms of square and purpose of *Mandala* is to harmonize the living space with the natural balance of the cosmos" (Pegrum 2002). *Vastu-Mandala* is known as a ritualistic geometry design and associated with the five elements of universe which provides us the basic knowledge of architecture that everything should be constructed in accordance with the right direction and proportion. The geometric figure of *Vastu-Mandala* deals with the metaphysical aspect of life. According to *Vastu-Mandala*, the house should have a perfectly balanced environment which enhances health, prosperity, and happiness in life. In this context, Hindu craftsmen had keen observations and were fully aware about *Vastu-Shastra*. Therefore they constructed the Hindu temples as per *Vastu-Shastra* norms. Along with this, geometry can be seen in the Indian Tantric art which is the combination of geometry and art. In *Tantric* art, geometry is a visual representation of deities as *yantras*. A *yantra* is based on basic primal geometric shapes with an abstract metaphysical thought. These geometrical shapes are psychological signs which provide a spiritual significance and higher levels of consciousness. *Yantras* are usually designed for meditation and religious rituals of *Tantric* art in Hinduism. In this regard, one such artist of India who was inspired by Tantric art and geometrical abstraction art is Ghulam Rasool Santosh who had created his own style of paintings which are recognized for spirituality, sensuousness, and neatness of lines and are influenced by Tantric art and Kashmir *Shaivism*. Another painter is Syed Haider Raza who was inspired by Tantric art as well as geometrical abstraction. His paintings are a visual representation of spirituality, energy, and creativity. Literature reveals that most of the contemporary artists were working in geometry abstraction art.

18.5.2 Application of Geometry in Islamic Art and Architecture

Islamic art and architecture demonstrate fascinating phenomenon of combination of intricate designs of geometry as ornamentation. The sacred architecture is the representation of the cosmos with highly rich divine symbols or words (Dua 2006). In this regard, Islamic architecture has generated a unity of purpose and principles in all manifestations, which bind them together. Due to prohibition of using living things in Islamic art, geometrical patterns have been created and developed by artists with Islamic philosophy, and they are lavishly implemented as a major design element in architecture. In India, during the Sultanate period, a wide range of intricate geometrical designs were introduced with religious faith (Nath 1976). For instance, geometrical patterns have been used in tombs of the Sayyids and Lodhis in Delhi. Nevertheless, huge numbers of varieties with complexity of geometrical motifs have been adopted in Mughal architecture as a main design element.

18.6 Arrangement of Geometry Patterns in Old *Havelis* of Saharanpur as per the Theory of Design Principles

A basic knowledge of geometry is indispensible in creating a driving force of innovative structural morphology (Cruz 2013). Geometry is associated as a science with technical support in architecture which generates the relation between geometry and architectural space through design principles and delivers the information about the application of geometry in architecture. It has the potential to develop innovative and sophisticated forms. In this regard, aesthetics and geometry have a correlation, because aesthetic perfection can be achieved through the accurate application of geometry. The purpose of the study is to describe and explain the arrangement of geometrical patterns in old *havelis* of Saharanpur as per the visual design principles of arts. Geometrical patterns of these *havelis* are beautifully organized and thoroughly and technically justified which lay emphasis on lyrical and aesthetical abstraction of forms and create a melodious atmosphere. In this regard, principles of art, such as unity of harmony, importance of balance, rhythm, and repetitions, play an essential part in the placement of geometry on the old *havelis* of Saharanpur.

18.6.1 Unity/Harmony

Unity is the focal component which lays emphasis on all elements of objects, and if combined together, these create balance and harmony in the architectural edifice. Unity is associated with a sense of completeness of all design elements. It can be categorized into three kinds: unity in static and dynamic, style, and theme. This can be seen in extremely exceptional carvings on wood as well as yellow stone in old *havelis*. In the *haveli* of *chhatta* Barumal, geometrical motif has been applied for filling the surface inside the stone brackets and creates a unity in theme with a variation of floral motifs (Fig. 18.10). The carvers very skillfully controlled the elements combined together and presented a beautiful composition of geometrical ornamentation in *havelis* which can be clearly visible in the *haveli* of Aatma Ram and few others (Fig. 18.11). It can be noticed that all patterns are different to each other but combined together will create unity (Figs. 18.10 and 18.11).

Fig. 18.10 Geometrical patterns on *haveli* in *chhatta* Barumal

Fig. 18.11 Geometrical *jali* patterns on corner in *haveli* of Aatma Ram (Source: Author)

Fig. 18.12 Wooden door of old *haveli* in *chhatta* Jambudas with layout and details (Source: Author)

18.6.2 Repetition

The repetition of design generates unity within the geometry design and makes an integrated effect which enhances the harmony in overall composition. This can be traced in well-proportioned designs and interchanging of sequence of designs in these *haveli* decorations. The character of repetition can be seen in wooden doors of *havelis* in *chhatta* Jambudas, and it generates rhythm, unity, and interest in geometrical motifs (Fig. 18.12). Repetition plays a key role in design elements in old *havelis* of Saharanpur because it produces a design free from tension.

Fig. 18.13 Geometrical motifs inside the stone brackets of an old *haveli* in *chhatta* Barumal (Source: Author)

18.6.3 Variety

Variety is the major character of design element which throws light on miscellaneousness of design and increases curiosity and visual interest in design. In this regard, old *havelis* of Saharanpur demonstrate the variation of geometric patterns which attract the spectators and denote the aesthetical unification. The variety in design has the power to gain the spectator's attention (Sharma 2010). The carvers of Saharanpur were fully aware of this interesting principle. Therefore, they made these geometrical patterns with intricate manner and variation of theme and used different motifs in one piece of art. For instance, it can be traced in *haveli* at *chhatta* Barumal which demonstrates the large variety of geometrical patterns inside the stone brackets. All brackets are carved out with divergent patterns of geometry that create a harmonious blending with other decorative elements (Fig. 18.13).

18.6.4 Rhythm

Rhythm is the combination of repeated patterns with variations. Through variation and repetition, rhythm is developed in artifacts which create a different feeling of music or dance. It is noticed mainly in geometric patterns of old *havelis*, which express exquisiteness through their schematic quality. Old *havelis* of Saharanpur display a movement of life through regular arrangement of geometrical designs which was achieved by individuals as a rhythmic sense. The carvers of Saharanpur applied the value of rhythm in geometrical patterns and increase the beauty of *havelis*. Therefore, the *haveli* of Aatma Ram is a visual manifestation of rhythmic unity through repeated patterns of octagon grid which mesmerize the spectators (Fig. 18.14).

Fig. 18.14 Wooden door of old *haveli* of Aatma Ram with layout and details (Source: Author)

18.6.5 Balance

"Balance is the most important characteristic of visual principle that a design should be weighted equally balanced on both side of a composition" (Hashimoto 2004). Balance can be classified into two ways: symmetrical and asymmetrical balance. In symmetrical balance, both sides of artifacts are divided into similar parts. Visual weight can be divided into the similar side of artifacts. A balanced design provides a feeling of stability. In an asymmetrical balance, both sides of artifacts are different but still look well adjusted. "An asymmetrical composition has two sides that do not match, but the composition appears to be balanced because the visual weights in the two sides are very similar" (Malloy 2014). The design elements of old *havelis* of Saharanpur are enriched with balancing action which interestingly increases the grace and charm through symmetry and asymmetry in geometrical forms. In this regard, symmetrical balance has been discovered in the *haveli* of Nanda Ghati and *haveli* in *chhatta* Jambudas. These *havelis* can be noticed for similar patterns through open lattices which increase the beauty of cusped arches (Figs. 18.15 and 18.16). Hence, the overall effects of geometrical patterns of these *havelis* are complementary and admirable with balanced forms.

18.6.6 Proportion/Scale

Proportion deals with unity of an object that creates a sense of completeness and relates all parts joined together. In a broader sense, proportion indicates the measurement or size of an object. Everything should be designed with the right proportions. In this regard, the geometrical images of old *havelis* have been created with technical skill and right proportion that hold the viewer's attention as well as give us immense pleasure. This principle generates the spiritual and metaphysical aura through the visual representation of design on surface.

Fig. 18.15 Symmetrical balances through geometrical patters of *havelis* of Nanda Ghati

Fig. 18.16 Symmetrical balances through geometrical patters of *havelis* in *chhatta* Jambudas (Source: Author)

18.7 Role of Geometry as Symbolic Tool in Old *Havelis* of Saharanpur

The significance of geometrical motifs in old *havelis* of Saharanpur is comprehensively connected with the symbolic aspect of universal expression, because these simple and complex forms of geometry reveal the vast reality of the motifs' hidden meaning which directly and indirectly influenced human life. "Art is prosaic without the delightful presence of sign and symbol. As well as, the spirit of fascination has also been generated through the symbolic manifestation of artifacts" (Verma and Gupta 2015). For instance, a triangle has three equal sides which directly represent geometrical properties like exact measurement, calculation, and logical reasoning, and on the contrary, the basic shape of a triangle expresses trinity such as in a

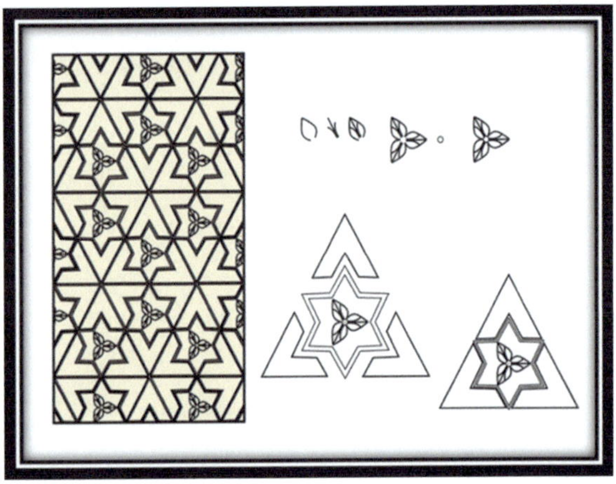

Fig. 18.17 Triangle pattern inside the bracket in the *haveli* of *chhatta* Barumal

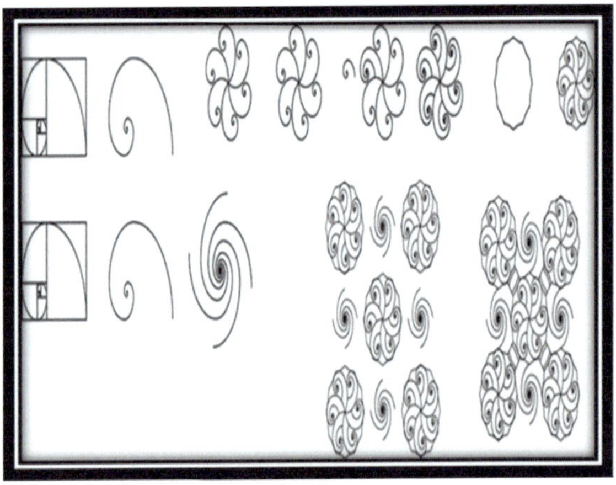

Fig. 18.18 Golden spiral pattern inside the bracket in the *haveli* of *chhatta* Barumal (Source: Author)

world where a human is associated with a god by worshiping. Worshiping is the path of that place where a human feels divine experience, in which he meets the supreme power of the universe. Therefore, a triangle may be noticed inside the stone brackets of the old *haveli* of *chhatta* Barumal as a symbol of trinity of the spiritual world and creates mesmerizing effects on the structural surface (Figs. 18.17 and 18.18). Spiral is the most profound pattern in Islamic architecture which represents the progress of the universe. "Spiral is a process of coming to the same point again and again, but at a different level, so that everything is seen in a new light" (Saraswat 2012). Spiral is a visual representation of the development of a human being. In this context, old *havelis* of Saharanpur display a variation of geometrical shapes and patterns like hexagon, spiral, rectangle, six-pointed star, and some grid patterns which have been developed with repetition, symmetry, and continuous generation of design. These geometrical patterns reveal a mystery of symbolism and give direction to human life.

18.8 Discussion

It has been mentioned above that the old *havelis* of Saharanpur are embellished with the idea of ornamental texicon (patterns, motifs, design) as well as the basic urge of an individual and demonstrate a profound knowledge of design elements. Now the question arises on why the design principles were applied in geometrical patterns and how these patterns enhance the beauty of these old *havelis*. Through the site survey, the study found that the carvers had techniques and technical minds too. For this, they followed some instructions and create laws for achieving aesthetic perfection. The term aesthetic perfection alludes to an excellence of geometrical patterns of old *havelis* in which all patterns have combined together and are well arranged which describes the specific configuration of each design. The size and arrangement of geometrical patterns of the old *havelis* were fixed according to the overall geometrical calculation. If a structure or pattern is made according to geometrical scheme, then it will reflect more beauty and will create a harmonious ambience. Firstly, justification offered the approach and the appreciation toward aesthetics delight of geometrical patterns of the old *havelis* and that those symbolic meanings are perceived by observers. The relationship between an observer and a design is often represented as a process of communication. The design elements have always been supposed to be a part of the sign system with which the observer constructs meaning. The communicative ideas of geometrical pattern of the old *havelis* are very strong which interpret the underlying meaning of patterns. The other point the authors would like to emphasize is the social issue related to geometrical patterns of old *havelis*. The social issue deals with the changes and continuity in geometrical patterns of the old *havelis* of Saharanpur. Carving is the tradition which is handed over from generation to generation. In this regard, geometrical patterns of the old *havelis* have been changed and transformed from generation to generation. Modification, variation, and adaption are the essential part of this tradition which indicates transformation. With the passage of time, these patterns are continuously used with transformation and modification by the carvers. It can be seen in the wooden folding screen which is not only a functional device but also a decorative craft. In contemporary time, folding screens are being experimented with modified geometrical patterns as per the requirement of user. The other point that needs emphasis is the sociocultural issue which deals with the companionship of two cultures. In this context, the supreme structural surface of the Islamic architecture and Mughal architecture attracts the carvers to take the inspiration for pure geometrical forms. These *havelis* are the glorious symbol of the Hindu-Muslim brotherhood and friendliness, because most of the *havelis* belong to Jain Hindu families who ornamented and painted their structural part with the amalgamation of Hindu and Islamic elements. Such geometrical and floral motifs are mostly profound in Mughal art and architecture, whereas figurative and auspicious patterns are the main characteristics of Hindu art and architecture. Therefore, geometry is the major constituent which can be seen as the structural part of Hindu and Muslim houses. "It may be possible to say that geometrical and floral decoration would be done by Muslim *ustads* (master craftsmen) and figurative and auspicious motifs would be done by Hindu carvers and painters" (Jain 2011). Indeed, these *havelis* and Jain temples are the best surviving example for this. Thus, it can be said that geometrical patterns have been applied in Jain *havelis* of Saharanpur which create a massive path of companionship, affection, and unity between Hindu and Muslim religion. Therefore, geometrical patterns played a very significant role in society.

18.9 Conclusion

Design element enhances the beauty of architectural space as an ornamentation tool which has the most powerful and admirable visual impression of human life. The decorative aspect of architecture is developed with the design principle as well as aesthetic theory of attractiveness. In this way, the geometrical patterns of old *havelis* establish the supremacy of proportions of harmonious approach and expressiveness of unity which is the major component of design element. In aesthetical and lyrical sense, these geometrical patterns of these *havelis* lay emphasis on inner beauty of visual image of geometrical patterns and denote a state of completeness and purity in which a miscellaneous range of aesthetical and design concepts combine together and enhance the attractiveness of *havelis*. During the survey of these *havelis*, it has been found that geometrical patterns in old *havelis* at *chhatta* Jambudas and Nanda Ghati are sharing similar patterns which create aesthetic sense. Variation and repetition of patterns can be noticed in the old *haveli* of Aatma Ram and *haveli* in *chhatta* Barumal and Jambudas. The concept of aesthetic perfection is applied through geometrical motifs in old *havelis* of Saharanpur because these *havelis* have been established with supreme characterization of grace. The placement and proportion of geometrical motifs of *havelis* are well connected with the variation of theme, and they are perfectly balanced with the structural surface. In current scenario, geometrical patterns of old *havelis* of Saharanpur have become a part of fascination and source of inspiration which are admirably being applied on woodcraft of Saharanpur. Carvers are inspired from geometrical patterns of these *havelis* and present geometrical design with innovation on craft items. For instance, wooden folding screens, tables, coasters, furniture, photo frames, and few others are created with geometrical patterns on the demand of buyers. They are creating interesting geometrical patterns with aesthetic perfection which is also a way to attract people after using such spiritual aspects. Maybe the placement of such designs is only a way to express Indo-Islamic culture of art and craft. Hence, the carvers of Saharanpur's woodcraft are deeply rooted with myth and religious attitude of Indian and Islamic culture. Another thing is related with aesthetical approach because in art numerous applications are done for marketing purposes. Therefore, the artistic mind is contented with many tools regarding the mythology, culture, moral values, and beautification too.

Glossary

Chhatta *Chhatta* is covered with a wooden overbridge that links two *havelis* of the same owner situated across the street without hindering the path below.

Geometrical Motifs Patterns composed of geometrical elements: square, rectangle, and stars with straight or curved lines.

Haveli A huge mansion.

Jagir Governor or ruler.

Lyrical Ambience Melodious atmosphere.

Ornamental Texicon Motifs, patterns, and symbols.

***Tantric* Art** *Tantric* art is associated with meditation as *yantra* form.

Ustad Master craftsmen.

Vastu-Mandala *Vastu-Mandala* is a symbolic geometric diagram used in architecture.

Yantra *Yantra* is a symbolic representation of deities as geometrical forms.

References

(SWCA), Saharanpur Wood Carving Mfg. Association, Retrieved 18 June 2015 from http://www.woodcarvinginindia.org/History.html

Cruz JP (2013) Structures and architecture: new concepts, applications and challenges. CRC Press

Dua S (2006) Relevance of geometrical systems in contemporary architecture (doctoral thesis), Indian Institute of Technology Roorkee

Hāṇḍā O, Jain M (2000).Wood handicraft: a study of its origin and development in Saharanpur. Indus Publishing

Hashimoto A (2004) Visual design fundamentals: a digital approach. Charles River Media

Jain DM (2011) Jain wall painting in Doab Region: with special reference to Saharanpur and Adjoining Areas. Indus Publishing Company

Malloy K (2014) The art of theatrical design: elements of visual composition, methods, and practice. CRC Press

Nath R (1976) History of decorative art in Mughal Architecture. Motilal Banarsidass

Pegrum J (2002) Vastu Vidya: harmonize your living spaces through the practical Indian tradition of Vastu. Duncan Baird Publishers

Saraswat S (2012) Spiral: a representation of process and growth. INSITE, a periodical of Institute of Indian Interior Designers (IIID). http://www.insiteindia.in/2012/may/insite%20addons.pdf

Sharma P (2010) A critical analysis of Inlay designs in the Mughal Monument of Agra (doctoral thesis), Indian Institute of Technology Roorkee

Verma A, Gupta I (2015) Indian Rasa manifestation in the motifs of Saharanpur woodcraft. Am Int J Res Humanit Arts Soc Sci 9(1):17

Chapter 19
Assessment of Mughal Mural Decoration on Contemporary Architecture of Agra and Jaipur

Rohita Sharma and Ila Gupta

Abstract Wall decoration and architecture have strong relations with each other. This relationship has been seen during the evaluation of wall decorations in the Mughal period and contemporary period. The Mughal period covers the fifteenth century up to the seventeenth century, and the contemporary period belongs to the twentieth century. There is a gap between the two periods. In present time, the concept, theme, and style have been affected by Western culture. The artist, without innovative ideas of Westernization or modernization in art, is being neglected by the society. At present, among the other changes, the impulses have dispelled from natural representation. The impact of abstraction may be seen in art forms, which has been broken by the tendency of making the natural things and converting them into abstract. This is the main theme for decorating a building in the present scenario. Except from all innovations, Indian art has its roots into its cultural and religious diameter. The present paper is an analytical study to explore the trend and trendies of present-time mural decoration in Jaipur architecture. This paper will also focus on the impact of Mughal art on contemporary wall decorations and also make clear if contemporary wall decorations have any impact from Mughal wall decorations or not.

The study is based on primary as well as secondary data. The sources of primary data are the many Mughal monuments of Agra, and for contemporary, these are the Dayal Bagh Temple and Kalakriti Emporium in Agra, hotels, and Vidhan Sabha Bhawan in Jaipur. The result has been finalized by discussions with artists, masons, historians, architects, and art conservators.

Keywords Mural • Mughal • Decoration • Contemporary

R. Sharma (✉)
Lovely Professional University, Phagwara, India
e-mail: rohitasharma61@gmail.com

I. Gupta
Department of Architecture & Planning, Indian Institute of Technology Roorkee, Roorkee 247667, Uttarakhand, India

F. Seta et al. (eds.), *Understanding Built Environment*, Springer Transactions in Civil and Environmental Engineering, DOI 10.1007/978-981-10-2138-1_19

19.1 Introduction

Mural or wall decoration is a part of architecture, which is an ancient art of the world. It survives everywhere today in the form of large-scale colored drawings on the walls of several edifices. Due to climate, atmosphere, and availability of material, these murals vary in style in different parts of the country. However, this art is always vigorous and full of movement. Europe, Asia, the USA, India, and Mexico are world famous for their beautiful mural decorations.

Mural art has great power in itself. It shows every phase of society. If anybody wants to know about the culture and living style of a society, then art is a strong weapon. It is same with this present research. This study will show the style, theme, and many other aspects.

In Latin language, mural = murus which means related to the wall. It is a part of art. Mural is a technique of art where an artist paints directly on a wall, a ceiling, and other permanent surfaces. "A mural is an extremely large work of art most often applied to the surface of a wall, ceiling, or floor for aesthetic and didactic purposes" (Wisegeek 2012). "It may be applied to any type of hard surface or on canvas and later fix it on wall frame. They may be executed on wooden panels and fixed on walls or painted on canvas in the artist's studio and glued to walls or ceilings, interiors and exteriors of buildings" (Howard and Opoku 2012). According to Adams (Adams 1990), murals and frescos are the same thing. Fresco is the important part of a mural. Tarantino (2011) explains that "Any tile or fired clay that is not glazed with a design or a number of such tiles that are individual segments of a larger design and fixed to a wall or floor can be referred to as a mural." Murals are not only related to the decoration only but also a medium to show something, as in the prehistoric period. It was a medium to present daily life activities, in ancient times to display symbols from religion and tell the stories of gods and goddesses, in medieval times to present a court scene, and in modern times to advertise and to get popularity as well as to decorate a plain wall and break the boring ambience by creating an illusion. Public commissions of murals in schools, hospitals, and retirement homes can achieve a pleasant and welcoming atmosphere in these caring institutions. The murals have varied due to its technique. "Mural technique includes encaustic painting, fresco, oil painting and tempera. Some modern murals are made of photograph and mosaic. Although fresco is the most commonly used technique, but there are other techniques of Mural art. Mural based on fabric and textile techniques seem to have limited application probably because of their susceptibility to deterioration with age" (Howard and Asare 2012). In this paper, the author will explore mosaic, inlay, and painted mural in Mughal monuments.

19.2 Mural in Mughal Monuments

In the Mughal era, different types of mural style have been used. They are very attractive and have an important place in the society. The most popular mural styles are mosaic, inlay, tile, stucco, carving, and painting. Inlay is one of the most popular of them.

19.2.1 Mosaic

A mosaic is a picture made of many tiny pieces of one and different colors (Fig. 19.1). "By mosaic we understand the art of putting together pieces of various materials either white or parti-colored in such way as to form definite patterns" (Birdwood 1880). "The Greeks may have learnt the art from the ancient Mesopotamians" (Furnival 1904). This art became very popular in Europe, Asia Minor, Syria, Iraq, Persia, and India (Cresswe 1932). Stone was easily available in India, which helped this art to flourish more than glazed tiles. "In India the stone mosaic, from this very early stage had been channeled into two styles: one was the tessellated style wherein square or rectangular pieces of stones of different colors were assembled and arranged together) so as to form a pattern other was the inlay" (Nath 1970) (Figs. 19.2 and 19.3).

19.2.2 Inlay

Inlay decoration is very durable for wall decorations. Inlay may be an advanced mosaic technique, which might have started in Jahangir's period. It is long lasting since it is laid in the sockets of a hard base such as red sand stone, marble, or wood.

Fig. 19.1 Mosaic Mughal monuments Agra

Fig. 19.2 Mosaic at the tomb of Itimad-ud Daulah (Sources: photograph by author)

Fig. 19.3 Mosaic at Fatehpur Sikri

Semiprecious stones enhance its beauty. There is conflict as to the origin of this art; some believe it originated in Italy and others from India.

In the Jahangir period, two beautiful monuments, the tomb of Itimad-ud Daulah and the tomb of Akbar Sikandara, were decorated with inlay and mosaics. Gradually, this art, flourishing in the Mughal period, reached its highest level in the period of Shah Jahan. The most beautiful work of this period is the Taj Mahal. According to Smith (1901), the tomb of Itimad-ud Daulah had mosaics of colored marble and inlaid work, but after the Jahangir period, mosaic ornamentation changed entirely to inlaid decoration (Figs. 19.4, 19.5, and 19.6).

19.2.3 Painted Mural/Tempera (Paintings)

It is well known that the Mughals were fond of art. Akbar was one of them. Akbar was fond of painting, miniature and portraiture, as well as architectural decoration (Blochmann 2010). In the Mughal period, many beautiful murals had been

Fig. 19.4 Inlay, Jahangiri Mahal (Sources: photograph by author)

Fig. 19.5 Inlay, gateway, Taj Mahal (Sources: photograph by author)

Fig. 19.6 Inlay, entrance, Taj Mahal (Sources: photograph by author)

Fig. 19.7 Christian motif on the wall of (Mahal-i Khass) Fatehpur Sikri (Sources: photograph by author)

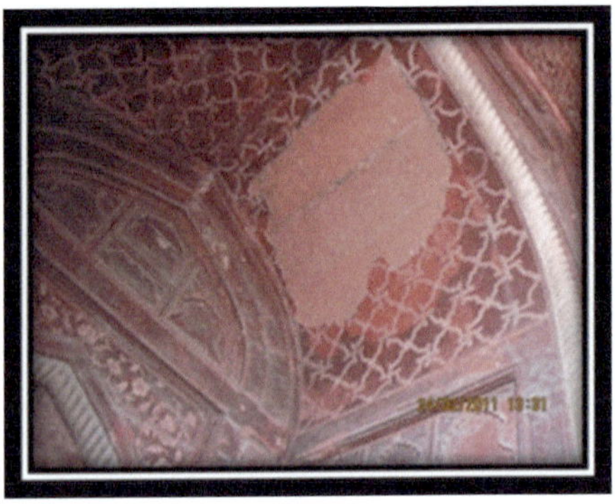

Fig. 19.8 Incised mural Taj Mahal (Sources: photograph by author)

decorated. Fresco-buono, Fresco-secco, and tempera are different techniques of wall mural paintings. The art of mural painting is a prehistoric art of India found on cave or rock walls. In this style, the artist paints with color on dry or wet plaster and sometimes only on a simple or plain wall. In the West and in Europe, many frescos have been built in palaces and temples of the ancient Greek and Roman civilizations. Persians were influenced by the European style, and from there this style was adopted by the Mughals. According to Annemarie (2004), murals were prominent features of castle and palaces. Akbar delivered ideas and the motive force, and the Islamic artist, chiefly Iranian, suggested the pattern as geometrical, floral, inscriptional, conventional, and also animate motifs, which were against the Muslim religion. Painted murals may be seen in the period of Akbar and Jahangir (Figs. 19.7, 19.8, and 19.9).

Fig. 19.9 Stucco with painted mural Sikandara (Sources: photograph by author)

19.3 Mural in Contemporary

As we are talking about the art in present time and come to its result with modernism, mural making in the contemporary era is an altogether different phenomenon, where religion is no longer a single or predominant subject. Contributing to an ongoing debate about modernity in India is vitally important. In the closing remarks of one of his lectures, Correa summed this up beautifully: "It will take time for the confidence to come back to a great civilization that lost every battle for 1,000 years. In India we have to reclaim modernity for ourselves. The whole world has a right to be modern, it's not a style, and it's a manner to present ourselves. At present mural is using to enhance the character of a building as British council library in New Delhi, and for a residential building it is like a punctuation."

With the exit of the Mughals, the art of mural work had started to decline. The number of craftsmen/artists engaged in this art began to diminish. In the mid-nineteenth century, there were only 100 craftsmen/artists specializing in this work and the art saw its revival in around the 1950s with the setting up of organizations like the Development Commission and the Handicrafts Board.

The study has been conducted to analyze which mural design elements and techniques are used in contemporary buildings.

19.4 Difference Between Mughal and Contemporary

To see the impact of Mughal mural decoration on contemporary mural decoration, it is necessary to know the other factors which can affect it. From Mughal to contemporary, lots of changes may be seen as contemporary designs depend upon user's taste and functionality of the building. Mughal designs are dependent upon rulers' taste and on traditional and classical themes. The motifs are more likely to be abstract and geometrical and stylized as casual and bold in present time, while in Mughal, they are more likely to be natural, floral, tree, plant, and landscape designs and geometrical motifs and religious/auspicious motifs such as *swastika*, *shatkona*, star, lotus, *purnakalasa*, and banyan tree. In modern times, the artists work in recently cultivated groups of artisans and designers/manufacturers, but in the Mughal era, artists were historically established and dependent on patrons or rulers. In this way, there are lots of changes. The factors which are responsible for this change are social factors, varying culture or religious perspectives, modern-style infrastructure, position of artists/muralist/interior designers, and political factors. For assessment of the mural in present time, the author made a selection of different public buildings.

19.5 Case Study

The assessment has been made through techniques, motifs, and themes on different case studies such as Dayal Bagh Temple in Agra, Kalakriti Emporium in Agra, hotels, and Vidhan Sabha Bhawan.

19.5.1 Dayal Bagh Temple in Agra

The Dayal Bagh Temple in Agra is the holy samadh decorated with flowers in inlay mural. The construction of this temple was started in 1904. This beautiful temple is being constructed in memory of the founder "Radha Soami Faith." The name Radha Soami was given out by the Supreme Creator himself (Farquhar 1924). "Samadh is a Hindi word. 'An immense marble and granite building is being constructed as both a place of worship and tomb'" (Juergensmeyer 1991). The whole building looks like a garden. The interior and the exterior walls of the building are decorated with flowers, plants, creepers, and trees by inlay and carving, which give the spontaneity. The entrance of the samadh is decorated by tree and flowers on the upper side. The side walls are also decorated with flower like kena, lotus, marigold, and many others. The main quality is that wherever we go we find the nature and greenery in the form of stones. The rose and the sunflower on the pillar come into view in waiting and welcoming for visitors. Every type of flowers, fruits, and creepers is decorated here in carving and inlay work. These flower motifs are in natural, decorative, and symbolic conditions. In Fig. 19.10, the geometrical and natural motifs are seen together, in the next figure is the subjective use of a flower, and the third figure is a motif of a flower vase with arabesque. These motifs, themes, and techniques are related to Mughal. The abstractive use of motifs, a flower as a subjective theme (Fig. 19.11), and a flower vase with wonderful arabesques are shown in Fig. 19.12.

Fig. 19.10 Geometrical and natural flower motif

19.5.2 Kalakriti Emporium in Agra

Fig. 19.11 Inlay decoration in natural flower (kena)

Fig. 19.12 Floor decoration in Dayal Bagh Temple in Agra

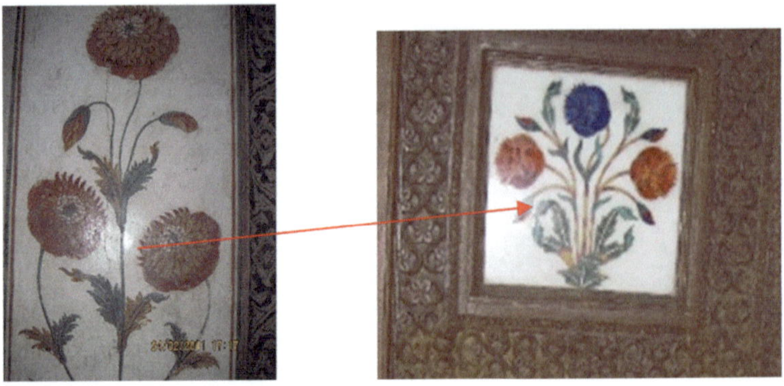

Figs. 19.13 and 19.14 Flower motif in the tomb of Salim Chishti (**Fig. 19.13**). Same motif in Kalakriti Emporium (**Fig. 19.14**)

Kalakriti is the largest and most spectacular handicraft emporium in Agra. It stands over 10,000 m^2. It is situated on the way to the Taj Mahal. The emporium is very close from the Taj Mahal on VIP Road, Agra. The building has an amazing entrance door studded with semiprecious stones and with a design that resembles the famous Buland Darwaza. Intricate carvings and patterns on the wall and door deem the embellished beauty of Kalakriti, which resembles Mughal beauty and takes us in their memories. There is some beautiful example which has been used in Mughal monuments. The interiors of Kalakriti are sprinkled with intricate wood and metal carvings. Each art effect is a timeless possession to beautify space. Many motifs may be seen through Mughal mural decoration as in the present figure the elements have been taken from the tomb of Itimad-ud Daulah and Taj Mahal and used in contemporary building. The similarities may be seen through these figures. A flower decoration (Fig. 19.13). decorated in the tomb of Salim Chishti may be seen on the door of Kalakriti Emporium (Fig. 19.14). The style and pattern are exactly the same. In others (Fig. 19.15), an ornamental flower vase with a lotus flower has been decorated in Taj Mahal. The same pattern is presenting on the door of Kalakriti Emporium. In Figs. 19.16 and 19.17, the lotus leaf has been used in repetition theme to create a wonderful effect. The same leaf has been used in floor design in interiors of the Kala Bhawan Agra. There are so many things decorated with inlay, glazed tile, and painting in that building.

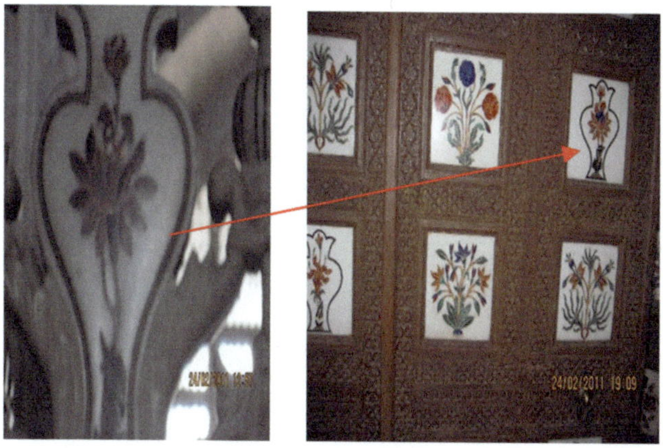

Figs. 19.15 and 19.16 Flower motif in Taj Mahal (**Fig. 19.15**). Same motif in Kalakriti Emporium (**Fig. 19.16**)

19.5.3 Hotels in Jaipur

Jaipur is a tourist place. Many visitors come across in India to see the culture of India. The hotels are decorated with different types of wall decoration having influenced by Mughals. The decoration may be seen in tempera, stucco, and mosaic. In present (Fig. 19.17), splendid decorative flower decorations have the resemblance to Mughal era flower decoration. Next (Figs. 19.18 and 19.19) shows the decoration of glazed tile which is a Persian technique and came under the Mughal era.

After visiting the many restaurants and hotels, the author found that most designs are inspired by geometry and abstraction. In present time, the motifs are used to attract every type of viewer. In other words, the motifs should be universally accepted. In Fig. 19.20, there is a *shivlinga*. The foreground and background are divided in different geometrical motifs, which are creating abstraction in this picture. In modern times, abstract is used everywhere in painting, design, and banners also. According to many artists, abstract is the main theme of modern art. This modernity placed the art of India equal to the same platform with other countries. But in the author's view, Mughals were modern in their time, because they started this modernity (Fig. 19.21). This miniature can show that the abstraction was present in Mughal miniature. This miniature is divided in different geometrical forms, which is showing different ceremonies in the inner as well as outer palace.

Fig. 19.17 Motif in Taj Mahal and in Kalakriti Emporium

Fig. 19.18 Decorative flower decoration in interior of moon palace hotel (Source www.google.commotifs)

19.5.4 Vidhan Sabha Bhawan Jaipur

From 1952 to 2000, the Sawai Man Singh Town Hall was being used for the Rajasthan Legislative Assembly. The 5th session of the 11th legislative assembly was the last session, which was held in Sawai Man Singh Town Hall on November 6, 2000. Work on this project commenced in November 1994 and completed in March 2001.

Fig. 19.19 Glazed tile decoration on the exterior lower part of the wall (Source www.google.commotifs)

Fig. 19.20 Abstract background in mural of a restaurant

Fig. 19.21 Abstract background in Mughal miniature

Fig. 19.22 Copy of mural of Amer Palace on Vidhan Sabha Bhawan in Jaipur (Sources: photograph by author)

The outer part of the building has been decorated with famous and decorative features of Rajasthani folk culture. For this reason, the Rajasthani art is different from other art, such as *jharokas*, *chhatries*, *kamani*, *baradaries*, *arches*, *todies*, etc. The interior portion has been decorated in the famous Rajasthani traditional art on walls as seen in different places of Jaipur, Shekhawati, Marwar, and Mewar. This is the modern-style architecture with many modern facilities. In a survey with an artist in Jaipur, he told us that the outer façade of the building is decorated with beauty – fully tempera and other decorated art as glass mosaic. The artist said that this building has an impact of Mughal era decoration (Fig. 19.22). The arrows show different motifs from Mughal mural designs on Vidhan Sabha.

19.6 Analysis

The result shows that flower motifs, abstract, and plant motifs are being used in most of the places. These motifs are universal motifs and accepted by 69 % of the respondents. So except from all the motifs of the Mughal era, these motifs may be seen in almost every space. In using this technique, most of the respondents give a positive response. The most preferable technique in present time is painting. The stucco is also used for sportive medium. Inlay is very costly so every person is not able to use this technique. Glazed tile decoration is now used in blue pottery, and many artists are working on it. But still most places are decorated with tile decoration. So it will be right to say that painting may be used in a high scale and by every common person also (Figs. 19.23, 19.24 and 19.25).

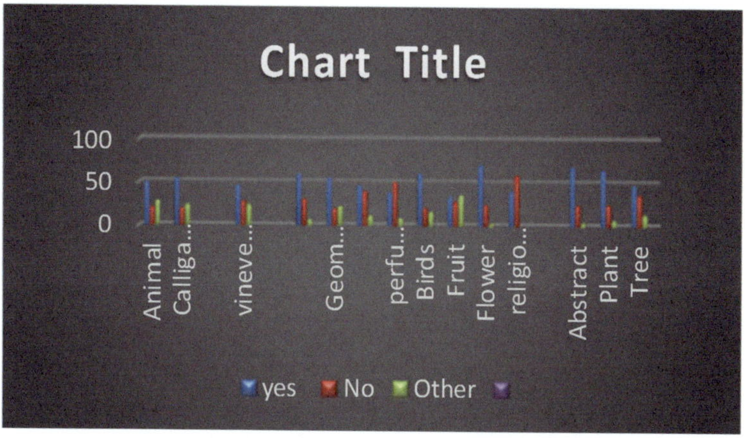

Fig. 19.23 Shows the most applicable mural motifs in present time

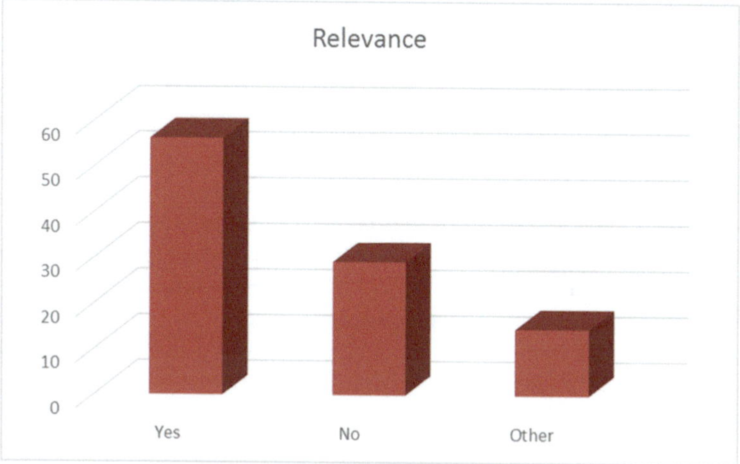

Fig. 19.24 Shows the relevance between Mughal and contemporary mural

19.7 Summary/Conclusion

Present time is the time of innovation by using new techniques and experiments. Realistic images of gods and goddesses are now converted into an impression through some symbols. So the copy of old murals continues with only some of the monuments,which may be present to show the legacy of Mughals or the culture of Mughals. Below are the examples of some old motifs as the copyright of Mughal and Rajasthani murals. In a interview of the artist of Vidhan Sabha Bhawan, he said that many motifs and techniques are equal to Mughal mural decoration.

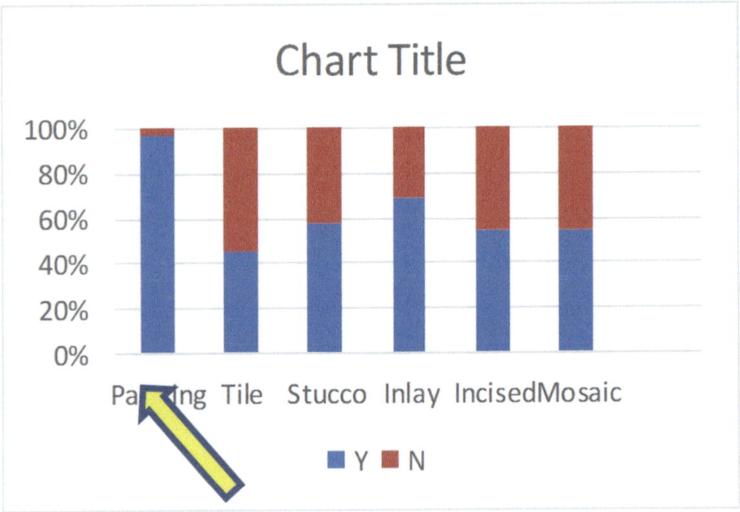

Fig. 19.25 Shows painting as the most preferable technique

To conclude, it may be seen that due to modernity, mural decoration has been changed. Technique of these decorations has added new material and equipment. The process which takes a long time now takes few minutes to complete. The themes which are used by the society in present have been changed. If we show it by a pyramid flowchart, then it will be opposite from the Mughal era.

Now there is a question: what is the reason behind this change? The answer may be the change in the living style of the society. In present time, the existing style is changed so the architecture for this changed living style is different from the past. Only for the different architectures, the style of murals is also different, but the origin of contemporary murals may be from the Mughal period. "Impact of Islam on Indian culture also had been inestimable as it permanently influenced the development of all areas of human endeavours – language, dress, cuisine, all arts and architectural forms and urban design etc." (Choudhury 1985). In this study, this has been found through surveys and interviews that mural decorations in present time are the status symbol. The people are using universal motifs.

- In present time, designers are known for featuring bold colors, abstract and geometrical shapes, and universal designs.
- In contemporary mural designs, a variety of traditional, modern, and universal trends are seen. However, geometrical designs are heavily preferred.
- It is noteworthy that hardly any difference is found regarding the placement of auspicious, geometrical, arabesque, and floral designs between the contemporary buildings and Mughal monuments.
- The designs and placement are according to the personal liking of the user, availability of ready-made designs, and type and nature of the building.

The purpose of murals in contemporary buildings is to act as visual communicator and as aesthetic enhancing element.

References

Adams LS (1990) Art across time – prehistoric to the fourteenth century, vol 1. McGraw-Hill, New York

Birdwood GCM (1880) The industrialist of India. London, p 212

Blochmann H (2010) Ain-i-Akbari of Abul Fazal. Vol. 1. Calcutta, p 107

Creswell KAC (1932) Early Muslim architecture, vol. I. Oxford, p 157

Farquhar JN (1924) Modern religious movement in India. J.S. CusingCo. Berwick & Smith Co, Norwood

Howard EK, Opoku NA (2012) Assessing the suitability of woven Fabric and composite textile techniques for mural production. J Sci Technol 32(1):140–149

Juergensmeyer M (ed) (1991) Radha Soami: reality the logic of a modern faith. Princeton University Press, Princeton, pp 15–20

WJ (1904) Leadless decorative tiles, faience, and mosaic, comprising notes and excerpts on the history, materials, manufacture & use of ornamental flooring tiles, ceramic mosaic, and decorative tiles and faience, with complete series of recipes for tile-bodies, and for leadless glazes and art-tile enamels. Stone: W. J. Furnival. Web

Makhan C. The Din-I-Ilahi, or, the religion of Akbar. New Delhi: Oriental Reprint, 1985. 1–47. Print

Nath R (1970) Colour decoration in Mughal architecture. pub. D. B.Taraporvala Bombay, pp 1–61

Annemarie S, Waghmar BK (2004) The empire of the Great Mughals: history, art and culture. Reaktion, London, Print

Smith EW. Mughal colour decoration of Agra: described and illustrated, ASI New Imperial series, vol.xxx, series 30 Allhabad, 1901 Print

Tarantino A (2011) Seattle street art – a visual time capsule beyond graffiti. http://en.wikipedia.org/wiki/Mural#Tile_mural. Retrieved 08/02/2013

Wisegeek what is mural? (2012) http://www.wisegeek.com/whatisMural. htm. Retrived 28/01/2013 stop karna hai red

Chapter 20
Gender and Space in the Paintings of Raja Ravi Varma and Amrita Sher-Gill

Mandakini, Ila Gupta, and P. Jha

Abstract The relation of gender and painting is a much debated issue not only in Indian art but also across the globe. In Indian concern of modern painting, this phenomenon has emerged as a critical expression. An artistic perception of onlooker/spectator and ideology of the artist on gender paradigm could be a contemporary topic of study. The space is selected as a criterion to analyze the gender discourse as feminine and masculine perception. In respect of modern Indian paintings, Raja Ravi Varma has been considered one among the first and foremost artists. He executed his paintings on Indian themes with Western techniques and portrayed numerous women with beautiful faces and sensuous appeal. On the contrary, Amrita Sher-Gill, also considered as one among pioneers of modern Indian art, has portrayed women in a different approach. In this way, both artists belong to modern Indian painters and contributed significantly with respect to art revivalism. Therefore, the depiction of women in both of the artists' works has been compared to explore gender significance vis-à-vis the painted space. A comparative analysis has been carried out between the paintings of Raja Ravi Varma and Amrita Sher-Gill through their stipulated space. The background space of the artists' paintings has a gender-inclined approach from masculine and feminine ideology. The significance of spaces has been evaluated on theoretical perception of some eminent art critics. Further, the background of paintings has been replaced through handmade paintings by using Photoshop as a tool. Thus, this paper is an attempt to evaluate the feminine approach in the works of both artists through the study of their negative and positive space in paintings. It concludes that the feminine characters placed in the interiors of places and household surroundings produce emphasis on gender study.

Keywords Painting • Gender politics • Feminism • Space

Mandakini (✉) • P. Jha
Department of Humanities and Social Sciences, Indian Institute of Technology Roorkee, Roorkee 247667, Uttarakhand, India
e-mail: mandakini.sharma9@gmail.com; Pashupatijha@yahoo.co.in

I. Gupta
Department of Architecture & Planning, Indian Institute of Technology Roorkee, Roorkee 247667, Uttarakhand, India
e-mail: ilafap@gmail.com

F. Seta et al. (eds.), *Understanding Built Environment*, Springer Transactions in Civil and Environmental Engineering, DOI 10.1007/978-981-10-2138-1_20

20.1 Introduction

The term gender has come into light during the 1890s for demanding equal status for women in all spheres. Gender in the architectural domain can be evaluated through space. Gender and space have been correlated with each other through various mediums including environment, objects, and background. In paintings, space is also equipped through negative and positive space and other supportive elements of theme and display. Raja Ravi Varma (1848–1906) has been called the father of modern Indian painting. He was also the first Indian artist who developed his own style with the combination of Indian mythical theme and Western technique (Parimoo 1998). In his paintings, he mostly depicted the sensuous and erotic woman in semi-naked condition. He molded the Western academic style in his own style of painting (Chawla 2010). On the contrary, Amrita Sher-Gill (1913–1941) was an artist who depicted the trauma and tragedy of womanhood through her modernistic approach (Wojtilla 1981). She was an Indo-Hungarian artist, who came up with feministic advancement. In both artists' paintings, space is an important aspect to come across the gender significance through a critical survey. So, the paper will be an attempt to discuss the role of gender in paintings through architectural space. Thus, some paintings of both the artists have been evaluated through space. The paintings of these artists have been compared after replacing its background through Photoshop software. These paintings have a powerful construction of gender politics in the portrayal of the feminine.

20.2 Review of Literature

Feminism encounters some issues, which work against the power or gendered metaphor through artistic representation in opposition to male-centered or idealistic connotations to question some political scenario regarding the portrayal of feminine (Kapur 2009). Rendell stated about the relation of gender and architecture; Pollock argued about the significance of space in paintings in respect of feminism (Rendell et al. 2003). According to Wojtilla (1981), Amrita Sher-Gill's description about feminine can be considered through her writings, in which the narration of gender difference and inequality in society can be found. She also portrayed such issues, including the "Child Bride." Women have been depicted in a sensual way by Ravi Varma; they also become the source of seductiveness and romance. Varma is the founder of calendar art in India and has been known as the father of modern Indian paintings. His numerous paintings are filled with Indian religious themes, but he has also portrayed his paintings on other subjects too (Thakurta 1991).

The historical representation of Indian art has been contented with gender politics through historical accounts, because it has degraded due to colonial encounters in India. This degradation has been done after the political manipulation of art narrations to decrease its originality by some non-Indian majorities, because they mentioned that Indian art is more feminine rather than masculine. On the other hand, some nationalists were also involved in this discourse due to their anti-European practices to build an authentic Hindu religion. And they are entirely against for such institutional practices, which were based on Western academic discourse. In this context, the Indian institutional practices cannot be ignored for reckoning the reduction of historical creations, in which the colonial perception was imposed for gaining supremacy (Garimella 1998). Aitken argued on the spectatorship and feminine concept in the miniature of Kangra style. The *nayika* or heroine has been portrayed with the group of women, which constructed the heroine as an object of gaze. In some paintings, the heroine has been constructed as an object of attention through the background settlements and the placement of figures (Aitken 1998). According to Lutzkar and Ann, through revolting against the patriarchal domination and gendered politics regarding the women status in India, the feminist artist, Amrita Sher-Gill, has inspired numerous female artists during the post-modernism, including Gogi Saroj Pal (Lutzker and Ann 2002).

20.3 Methodology

The study material of this paper has been collected from several resources. The collective stuff of the study has been classified into two categories, namely, primary sources and secondary sources. The secondary sources are commonly books, journals, and catalog of the shows (exhibitions) of artists. Those have been collected through library visits to the National Gallery of Modern Art (NGMA) and Lalit Kala Akademi in New Delhi. The other data and information have been collected through online sources from different websites. In the primary sources, there is one interview with Gogi Saroj Pal (contemporary artist) and one with Vinod Bhardwaj (an art critic) to evaluate the major difference between the paintings of male and female artists. In this procedure interview guide has been followed to know their perception.

Some paintings of Amrita Sher-Gill were collected by taking their photographs from NGMA with the permission of the director of the library. Then some of them are edited later for changing their background by using Photoshop software. Then these edited paintings have been painted by the author for the replacement of background.

During the interview with Gogi Saroj Pal, the author has found some approaches to understand the painting of Amrita Sher-Gill and to structure space and gender significance as compared to Raja Ravi Varma. Vinod Bhardwaj's ideas and views are also valuable to consider and evaluate such political issues regarding these paintings. Some scholarly writings are also observed to seek the significance of gender in terms of background and space in paintings.

20.4 Painting in Interior Decoration

Painting has an important place in interior designing. It has been used to decorate the surface or to communicate something to its onlookers. In this context, the first evidence in the history of painting has been found in the interior of caves. These cave paintings can be identified as a source of visual communication or as a medium of expression. The rock paintings of Indian prehistoric period have a remarkable approach with their strong lines, but they are more communicative rather than decorative. Afterward, murals of Ajanta have been painted for the promotion of religion with lyrical balance and wonderful gestures. Then this tradition of wall paintings has been further advanced as portable paintings, in which a piece of art can be patronized or bought by the individual. In this way, a piece of art or painting has been produced as per the taste of the individual and then collected or patronized by other persons. This taste is often differentiated as per the surrounding of the interior; for instance, it can be religious or commercial.

Painting is used to meet the ends of the consumer either to decorate the surface or to communicate their perception about something. Painting covers almost all spheres of realistic and imaginative world, which is dependent on the choice of the artist. It can be an individual entity or the universal one. In this larger domain of painting, architecture or interior space has also been implemented, which is mostly used to support the background. Moreover, in some landscape paintings, architectural buildings have been made directly as a subject as per the interest and style of the artist.

In current scenario, many prominent artists have come into picture, which is recognized through their individual style and application of theme. In terms of contemporary Indian architecture, these paintings are mostly displayed on the walls of interior surface of elite groups. The other groups, excluding the elite class, are also using some common painting, or reproduction of such paintings, as per their interest and financial status. Some emerging artists are also seeking their identity to come in such category. Therefore, the art of painting is closely connected with the interior decoration because it is not only used to decorate the surface but it has also its own narratives to represent various perceptions. Such perceptions can be related to religious, cultural, and numerous issues related to human interests. In these subjects, architectural space has been frequently depicted to give suitable surroundings to the subject or to create balance on the surface. In recent trends, painting has been become a major medium, and a valuable aspect too, in the interior decoration.

20.5 Space and Painting

Space is the most essential component of painting; this constructs the main objective of putting the subject and the objects on two-dimensional surfaces to produce three-dimensional forms and effects (Rendell et al. 2003). Painting constitutes of

Fig. 20.1 *Saraswati* 1896, oil on canvas, collection Maharaja Fatesingh Museum Baroda (Source: http://www.1st-art-gallery.com/thumbnail/233244/1/Goddess-Saraswathi.jpg)

two types of spaces; one is positive and the other one is negative. Positive space has been created through the figure or main structure or shape, and negative space belongs to the background and other secondary elements of painting. It is not only the main object that exhibits message of the painting, but suitable background represents the theme and idea of representation. In this context, the painting of the Goddess *Saraswati* can be observed to understand the relation between the space and painting (Fig. 20.1). *Saraswati* is made in open space, so the exterior of the painting has some architectural approach through space management.

In this painting, the Hindu Goddess *Saraswati* is depicted by Ravi Varma. This painting has been made for a religious purpose, where a Hindu Goddess has been depicted with a beautiful landscape.

If the painting is observed, it tells a story through its space making. The story of Hindu mythology regarding the Goddess *Saraswati* can be easily recalled through the negative space, in which the landscape is full of greenery. Deities are considered to reside in lonely places with gardens and ponds in Hindu tradition. Here, the pond has been made with lotuses because lotus is an auspicious symbol and often associated with the Goddess. The Goddess has been made as sitting on a rock and her vehicle (or *Vahana*) is also placed in a corner. As per the story, all these things associated with the Goddess create a sense of auspiciousness and devotion for its followers. This story is also acceptable for the onlookers because, as per mythological records, it is portrayed in a right manner. Ravi Varma has mainly been recognized for his calendar art, and this icon of the Goddess is also appropriate for that perspective.

Fig. 20.2 Edited image of Saraswati in Photoshop (Source: http://www.fellowshipofhope.net/wp-content/uploads/2010/12/DSC00183.jpg)

On the contrary, if the space of the painting has been replaced with some common background, the meaning of the painting has been changed automatically. Here, the importance of space construction can be considered through the replacement of negative space, which is replaced with a slum scene. In the next image (Fig. 20.2), the replaced background is full of uneasiness and disturbance because a majority of people, which differentiate such religious icons on those bases of supportive background, cannot be satisfied with this replacement. Visual perception of balance between the theme and background is representing an imbalance after this change. The religious appeal has been changed into awkwardness. Moreover, if the rock used for sitting will be replaced with a chair, the religious appeal will be turned into humor or irony. Therefore, space plays a major role to represent the ideology of the artist in respect of painting.

20.6 Gender and Space

The word gender has been derived from societal differences, which are not biological but directed through society on some cultural and traditional values. These values have been derived from a set of practices, in which masculine and feminine term has their certain discourse to perform with. This gendered aspiration has been raised as a political agenda of identity in respect of feminine. In the field of art and architecture, woman artists and architects have differentiated this discourse into a new paradigm, where they explore such issues, which were never considered before. These issues are not related to their physical beauty or sexuality, but to their self-exploration. They are more concerned about the subjective approach of womanhood rather than the objective one. This kind of opposition has led them to the academic discourse on feminism.

The quest of feminism is a refusal of feminist artists or a "discourse against power" for modernist art and an approach to renovate feministic brilliance through the subversion of gender politics and societal hegemony or "along such reconstitutive hypotheses that address the issue of power" (Kapur 2009). In paintings, such kind of gender politics has been explored by numerous art critics. In this process of exploration, the theme of representation is not only observed to differentiate such political issues, but their architectural background has also been equipped to represent woman as a thing to be displayed. Woman portrayal always remained a repetitive aspect in Indian paintings (Garimella 1998). In the earliest Indian paintings, a woman is often made to satisfy the male gaze or to represent the strength and heroism of manhood. These political representations can be seen through projection of the space, in which they are often depicted in a nonsignificant role, i.e., as waiting for their lover, engaged in erotic display, and involved in idealistic representation as per the social code of conduct.

Their own identity was mostly ignored, and they were portrayed as an object for the onlookers through space making. In such paintings, space is politicized with various aspects because, in the positive space, the woman is portrayed as a symbol of beauty and grace with sensuousness and the negative space is always copped up with such objectifications. In negative space, the surroundings of a woman have been made through the interior of a house or riverbanks, gardens, and so on. Moreover, male existence is always represented with female depiction, either they are physically shown or not shown, through symbolic significance (Aitken 1998). In such symbolic representations, space is a major phenomenon to observe this male domination in the representation of the feminine, for instance, toilet scenes, scenes of harem (a place where a bunch of females are particularly occupied with a male), bedroom spaces, places of roofs and gardens, etc.

In the Indian scenario, Ravi Varma has been considered the father of modern Indian paintings. He made numerous paintings on women, and his style was a combination of Indian mythological themes and the Western academic techniques (Ramachandran 2003). Woman representation in Ravi Varma's paintings is full of erotic display and offering (Verma 2003). First of all, in the matter of theme, they are characters of Indian idealistic module, and, secondly, they are serving themselves to its onlookers. The space of these paintings is occupied with male-dominated hegemony. Amrita Sher-Gill is the first female artist of India, who studied woman in a diverse angle and explored a new criteria in respect of womanhood, particularly in the last phase of her oeuvre (Dalmia 2014). She portrayed women in a realistic manner and created stress and uneasiness through color and background. In Sher-Gill's paintings, positive space is highlighted with the emptiness of background.

20.7 Discussion

In painting, interior space has been equipped to construct gendered dimensions by both the artists, in which Varma's women are full of objectiveness and Sher-Gill's women are full of subjective approach. In both of the artists' works, space is preferred in their own way.

In Ravi Varma's "Reclining Woman" (Fig. 20.3), the main female figure can be considered as positive space, and other elements are negative space. The Reclining Woman of Varma has attractive gestures and positions and a similar treatment like Edouard Manet's "Olympia" made in the nineteenth century (1863), which may be a result of Western influence. In this painting, the main figure has been made in erotic mood to seduce her onlookers. The background or negative space is constructed through an interior of a bedroom, which is also representing the state of waiting for her male counterpart. She is entirely conscious about her onlookers and staring into the eyes of them with a seductive approach. The ambience, which is created by the artist, represents a luxurious environment. The way she is laying on bed looks like a projection of the woman in front of the onlooker. The reclining figure of this painting has also been made in a sense of consciousness, i.e., she is aware about the gaze of the spectator or she is serving herself through an eye-catching sight.

The open book, which is used as a sign of the elite class (because literacy is also related to the upper class at that time), has no use in painting; it is rather a display. Another woman or attendant is painted in semi-naked condition, which may be a division of the elite and low class. But it can be considered that the main figure is covered fully, while the attendant has been made in semi-naked posture. Moreover, the central figure is looking more appealing rather than the second one due to the projection and gesture of the figures, because the central figure is quite communicating with the onlookers rather than the second one.

Fig. 20.3 Reclining Woman (undated) by Raja Ravi Varma, oil on canvas (Source: http://upload.wikimedia.org/wikipedia/commons/5/59/Raja_Ravi_Varma,_Reclining_Woman.jpg)

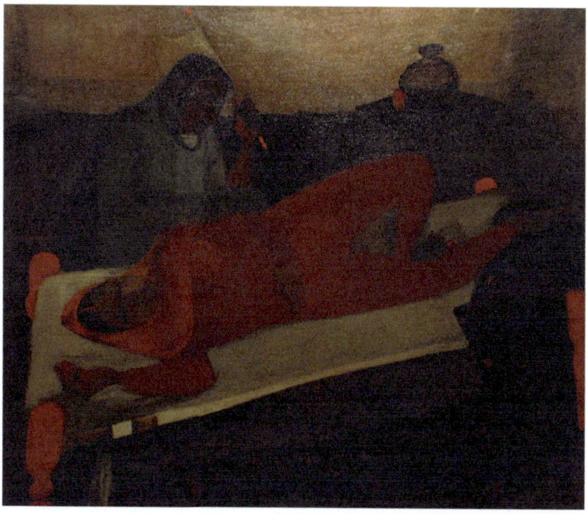

Fig. 20.4 Woman on Charpoy (1940) by Amrita Sher-Gill, oil on canvas, 33.46×28.5″ (By Author from NGMA)

Sher-Gill's Reclining Woman (Fig. 20.4) is entirely different from the Reclining Woman of Ravi Varma. In this painting, Sher-Gill has filled her woman with full-red garments and dark complexion, which is very rare during the Indian modern era. So, the positive space of Sher-Gill has a feministic appeal as she demonstrated their realistic aspect of life. The fully covered figures demonstrate the indigenous patriarchal hegemony with conceptual tension; they are not gazing into the eyes of the onlookers but are engaged in themselves (Kapur 2000). The background and placement of figures are in the interior of rural space of an Indian village. The negative space is also approaching its viewers, bringing attention to observe the absence of beauty and sensuousness and to explore their domestic sphere. Such kind of approach is missing in Varma's "Reclining Woman." Moreover, the theme of painting also originates through space making. For instance, Sher-Gill used *charpoy* instead of a bed, and the woman of Sher-Gill is laying in resting posture, which represents tiredness (Doctor 2002). Varma's woman is depicted in a seductive posture in a luxurious room (Thakurta 1991). Color application is also very dull in Sher-Gill's painting, which supports positive space. The other woman having a handmade fan does not look like a maid; the same stress can be caught on the faces of both the figures. The feeling of emptiness in painting is derived from loneliness of women, a common ailment among rural Indian women. The painting can also be a resemblance to the artist's life because Sher-Gill faced many grievances throughout her life (Chawla 2003). According to Geeta Kapur, the rural and aristocratic women of Sher-Gill have "Sher-Gill's peasant and aristocratic women in a feudal setting seemed to hold in their stillness a latency of desire that would appear to exceed the artist's own potential as a middle class women. But perhaps it was precisely the other way around: that the artist's (occidental/romantic, perhaps bisexual)" (Kapur 1978). The bright red used in painting is a symbol of marriage, and the red

Fig. 20.5 Edited in Photoshop and painted by author

sindoor filled on her head is a sign of her marital life, in which the women are filled with restriction. Sher-Gill's painting also looked like a satire for the erotic lover of women. The red four pillars of charpoy seem like a boundary which is constructed like the resemblance of a cage for married women.

In the next image (Fig. 20.5), the background of Sher-Gill's painting has been replaced with the background of Varma's painting, where the significance of negative space can be considered through placement of things. First of all, if the background will be observed through walls of the interior, Sher-Gill depicted a rural scene in her painting; the walls are painted in white with the mixture of dullness; on the contrary, Varma's wall has been made in brownish shade. Varma's wall is quite complementing the positive space because the woman is made in white *saree* and with brownish wall; she is appealing appropriately to its onlookers. In the replacement of background, Sher-Gill's painting is representing a misbalance of the surface with the central figure and theme, because brown is appearing very awkward with bright red. Even the couch is also creating a disturbance and does not match with the positive space. Sher-Gill has depicted an emptiness and stressed background, where only a small pot of water has been painted, which support the rural depiction, while after changing the background, the curtain is also not copping up with the theme because Sher-Gill depicted a marginalized woman, where the curtain and the couch are also not going with the theme.

The replacement of background has entirely changed the meaning of the painting in a different manner, and it appears as a misbalance including all objects. The attendant is also not matching with the status of central figure because the central one is fully covered and the other one is in a semi-naked condition. In the second replacement (Fig. 20.6), the background is again replaced with Sher-Gill's background, and here the importance of background and negative space may be explored instantly. From the very first site, the positive space or central figure is looking excluded from the negative space in color application, placement of objects, and so on. Varma has depicted a Nair woman with seductive appeal, in which its onlooker

Fig. 20.6 Edited in Photoshop and painted by author

encounters an elite class lady, reclining on her couch. On the contrary, this replacement (Varma's) is not communicating the aspect, which the artist wants to convey; for instance, as per the perception of John Berger, seeing and recognition come first before the interpretation of words or assumption regarding something (Berger 1972). Varma has also represented this painting to interpret his perception through the placement of background, but when the interior space has been changed, it has not only dismissed the position of positive space but has also converted the theme into an otherness of visual approach in all dimensions.

In another painting (Fig. 20.7), Ravi Varma has portrayed three women in a garden. This theme could be particularly related to the ancient Indian poetic story, *Abhijnanasakuntalam* of Kalidasa, a legendary poet of Sanskrit. In this image, Shakuntala has been portrayed with her attendants, so here are three female figures in three main postures. This kind of theme has been frequently depicted by many artists to delineate the structure of woman from every angle. This subject was first depicted by Botticelli (1445–1510) during the fifteenth century and then followed by Raphael (1483–1520) in the sixteenth century and many more Western artists. For instance, Picasso has also made paintings, in which three women have been made in etching.

According to Greek mythology, three graces are considered the goddesses of beauty, charm, and creativity. Ravi Varma was also inspired by the Western art, so he may take the theme and merge it with the Indian epical character named Shakuntala, who was a beautiful woman and known as a symbol of love and loyalty for her counterpart. In this image, Shakuntala and her attendants are positive space, and the forest and other surroundings are negative space. As per the theme of the painting, Shakuntala has been depicted as removing the thorn from her foot, but actually she is pretending to look at King Dushyanta. In this way, the feeling of first sight love or attraction has been symbolized through the thorn. The love of the king has pierced the heart of Shakuntala and this situation has been portrayed by the artists. So here, the woman is depicted as captured in the love of a man, and, more than

Fig. 20.7 Shakuntala (1870) by Ravi Varma, oil on canvas (Source: https://upload.wikimedia.org/wikipedia/commons/f/fa/Raja_Ravi_Varma_-_Mahabharata_-_Shakuntala.jpg)

that, he entirely dominated the painting in his absence. In this context, positive space is almost occupied by the male, in which a main character and her attendants are only placed to depict beauty and grace of the feminine. The negative space of painting is made to support the theme because the concept of the thorn is suitable in the forest according to the story.

On the contrary, the painting of Amrita Sher-Gill, titled "Group of Three Girls," can also be considered, in which three women have been portrayed to represent a different scenario of womanhood (Fig. 20.8). The theme of the painting is not related to any ancient story or mythological character, but it is related with a realistic aspect. The characters are not engaged in any erotic display or love act, but the women are displaying their tensed gesture. They do not belong to male-centered cultural hegemony, but they are particularly the women of artists' surroundings or their inner quests as well as they are the positive space of this painting. They are not displaying themselves as the onlooker wants, but they are displaying the actual state of their life. This statement has been given by Gogi Saroj Pal (a contemporary feminist artists) to the author during the interview.[1] Negative space is entirely blank

[1]This kind of observation is the prior concern of Sher-Gill in this painting because she spent most of her life in West and after coming to India she observed these subjects as like an outsider. She became an observer of these issues because they are uncommon to her. It is added by Gogi Saroj Pal, "Amrita was like a European painter and when she returned to India after her marriage she found Indian common people very attractive and uncommon. She painted rural Indian women because she never explored such aspects before in Hungry but she knew the structure of painting very well. She also explored an Indian sense to paint such realistic issues".

Fig. 20.8 Group of Three Girls (1935) by Amrita Sher-Gill, oil on canvas, 73.5 × 99.5 cm (By Author from NGMA)

instead of shades of positive space for giving entire attention to the women. So, going through the comparison of both of the paintings, their architectural space can be politically considered to frame out the ideology and perception in the painting. She is lovelorn for King Dushyanta because it's the time of her first encounter with the king.

In the above images (Fig. 20.9), the background has been replaced with one another. In this context, the first image (Ravi Varma's) has almost lost its identity. These figures are appearing as hanging on the surface; even their movement is also confusing. Their placement and balance have been deviated from their reality because the exterior space, through which the artist has used to give the story a romantic touch, has been gone entirely. The movement of background, which has been depicted by the artists through a bright background and greenish environment, has also been turned into stillness. This stillness has made positive space only a still-life kind of image rather than narrative. In Sher-Gill's painting (Fig. 20.10), the background full of greenery and other landscaping, in regard to the exterior space, is not much relevant as the previous one, because Sher-Gill has given more value to positive space and almost neglected the interior or exterior space. It may be a way to give more importance to the positive one, but, after the replacement, this direct approach of tragic representation has been lost.

Therefore, the space in Amrita Sher-Gill's paintings is entirely different than the painted space of Varma. There are certain reasons, which identify this very comparison on the bases of visual communication. In "Woman Resting on a Charpoy" and "Reclining Woman," there is only one similarity, which belongs to the theme of resting or lying, in which the main figure or positive space has been portrayed with

Fig. 20.9 Edited images in Photoshop

Fig. 20.10 Edited images in Photoshop

her attendant. And then, after going through the overall evaluation, it is observed that gender affects the works of both the artists because both artists have painted women in a very different manner. In Reclining Woman, the woman is portrayed as a medium for evoking sensual delight through beauty and delicacy, and the interior of the painting is also supported by the same aspect. As a male artist, beauty and sensuousness are preferred by Varma, while Sher-Gill has represented the reality of women's life with their strange faces. This depiction of reality by Sher-Gill has become more relevant because she avoided the interior details. She has focused only on positive space, so that woman remained the main concern to its onlookers. Varma preferred the beautiful woman and a sight of interior, which represents the same ideology, prevailed from the ancient times, in which a woman is considered as an object of erotic desire and often portrayed to satisfy the male gaze. On the other hand, Sher-Gill's women look like representatives of feministic quest, who are questioning this male gaze through their strange faces and rigid appearance. In "Shakuntala" and "Group of Three Girls," the comparison has been done on the basis of exterior spaces, where Varma has chosen greenish background and Sher-Gill has adopted the empty one. Varma preferred the mythological story in his own way, while Sher-Gill has focused on day-to-day life scene in a tensed ambience. Their respective backgrounds of paintings are entirely supporting to their ideology of representation, but this can be differentiating through the replacement of background. After replacing the background with each other, both artworks have lost their identity. The idea of portraying a beautiful woman, in a natural landscape, has been distorted through the empty background (of Varma's Shakuntala). The loneliness of Sher-Gill's icons is changed into other narratives, where sad women are sitting in a greenish background because a greenish background is not appropriate to depict sadness or tragedy. In Reclining Woman, the woman or positive space of the painting is particularly belonged to a seductive approach, and the negative one is also suitable for the theme. On the contrary, in "Woman Resting on a Charpoy," the woman does not belong to the elite class but is looking like a common rural woman. Through the placement of such surroundings, Sher-Gill has tried to capture the tough life of Indian rural woman. In that life, they have to do tedious jobs and then take rest for a short time. Varma has chosen only that part of a woman's life, which only belonged to physical representation for a male-dominated society, while Sher-Gill has chosen the part, which only belonged to a woman. After the replacement, "Reclining Woman" has become a piece of humor and irony, while "Woman Resting on a Charpoy" has become a piece of less consideration.

20.8 Conclusion

In conclusion, gender plays a major role in the construction of space. Gender difference may not be appropriately depicted in Varma's painting, but it is discovered in Sher-Gill's works. The women of Ravi Varma are placed in an ambience of rich/elite background and that of Amrita Sher-Gill's are surrounded in settings of sorrowful rural background. The woman, despite of similar posture of lying on bed, expresses all through the meanings of seduction and sadness. The negative spaces of interiors enhance these qualities of seduction and sorrow. The negative space when altered with a different atmosphere may not remain that appealing as it was before with the related one. The gender placement emphasizes their state of being that brings about different aspects of feminine in their architectural setting of background space. In Varma's negative space, the background and the interior have been taken as preferred aspects and equipped to represent the seductive approach. On the contrary, Sher-Gill gave preferences to positive space for women to communicate their tragic and painful stories. Male presence is always symbolized in Ravi Varma's numerous paintings, including the above-discussed paintings too. This symbolic significance can be traced through the interiors, surroundings, and gestures of the woman. Moreover, their absence is being equipped to politicize the space.

In Sher-Gill's paintings, the negative space has been mostly left blank and isolated because the protagonists of Sher-Gill have been portrayed to tell their stories, particularly belonged only to them. On the contrary, in Varma's painting, women have been politically manipulated to tell the story that is only associated with the objectification of womanhood. Therefore, the negative space of Ravi Varma's painting has lost its identity after the replacement of background; its main objective behind the portrayal has been entirely interrupted and become questionable in a strange way. The reclining woman in Varma's painting was portrayed to represent elite women with her attendant, and she has become only an object of humor and misbalance after the replacement of background. The first impression of the original painting entirely belongs to offering sensuous appeal, but the edited one is full of strangeness and not fitting into the edited background. Background or negative space of the painting has much importance as the positive one, and it helps to generate the identity of the painting. This projection of space varies in both artists' paintings.

On the other hand, Amrita Sher-Gill has represented women who have their own identity. They are also representing themselves for the onlookers, not for offering but for questioning the objectification of womanhood. In another painting, titled "Group of Three Girls," Sher-Gill has projected the emptiness of rural women with their strong gestures, while Varma's painting titled as "Shakuntala," portraying three women, has been associated with an ancient love story. Shakuntala has been depicted with two girls, where she is looking at her lover after being pretentious. Her lover is not depicted in the painting, but the painting's main objective may be an attempt to portray that lovelorn woman, who has no individual identity but acknowledged only to represent the manliness. In Sher-Gill's "Group of Three Girls," belonging to manliness has been neglected by the artist, and the strong political manipulation has been equipped for shifting this prevailed objectification. To conclude, it may be considered that the characters of Sher-Gill's paintings belong to a realistic world rather than any mythological or ancient fable, unlike that of Varma. The second strong characteristic of Sher-Gill's portrayal is her selection of sad and tragic faces rather than beauty and sensuousness. In respect of space, Sher-Gill's paintings are more focused on positive space because a woman is the only concern of the artist, who tells the story of her women herself, without any supportive elements.

References

Aitken ME (1998) Spectatorship and femininity in Kangra style painting. In: Dehejia V (ed) Representing the body – gender issues in Indian art. Kali for women – in association with the Book Review Literary Trust, New Delhi, pp 82–101

Berger J (1972) Ways of seeing. Penguin Books and British Broadcasting Corporation, London

Chawla R (2003) Talent, tragedy and the myth of Amrita Sher-Gil. In: Sinha G, Sinha G (eds) Indian art – an overview. Rupa. Co., New Delhi, pp 37–50

Chawla R (2010) Raja Ravi Varma – painter of colonial Inadia. Mapin Publication, New Delhi

Dalmia Y (2014) Amrita Sher-Gil: Art and life: a reader. Oxford University Press, New York

Doctor G (2002) Amrita sher Gil: a painter life. Motilal Books, London

Garimella A (1998) Engendering Indian art. In: Dehejia V, Dehejia V (eds) Representing the body – gender issues in Indian art. Kali For Women – In association with the Book Review Literary Trust, New Delhi, pp 22–41

Kapur G (1978) Modern Indian artists. Vikas Publishing House, New Delhi

Kapur G (2000) When was modernism: essays on contemporary cultural practice in India. Tulika Books, New Delhi

Kapur G (2009) Gender mobility through the lens of five women artists. In: Sinha G (ed) Art and visual culture in India – 1857–2007. Marg Publication, New Delhi, pp 300–319

Lutzker, M, Ann M (2002, November) Five artist from India. Woman's Art J, 21

Parimoo R (1998) The legacy of Raja Ravi Varma – the painter. Maharaja Fateh Singh Museum, Baroda

Ramachandran A (2003) Raja Ravi Varma – marketing strategies of a modern Indian artist. In: Sinha G (ed) Indian art – an overview. Rupa. Co., New Delhi, pp 23–36

Rendell J, Penner B, Borden I (2003) In: Rendell J, Penner B, Borden I (eds) Gender space architecture: an interdisciplinary introduction. Routledge in Taylor and Francis e – Library, London

Thakurta TG (1991) Women as 'calender art' icons: emergence of pictorial stereotype in colonial India. Econ Pol Wkly 26(43):91–99

Verma R (2003) Raja Ravi Varma – the pinter prince 1848–1906. Parsram Mangharam, Bangalore

Wojtilla G (1981) Amrita Sher-Gill and Hungary. Allied Publishers Private Limited, New Delhi

Chapter 21
The Idea of Infinite in Indian and Western Art: Perceiving It Through the Intangible Cultural Identity

Vandana Sehgal

Abstract The cultural identity of any society is unique and is shaped by its beliefs and ideas. These ideas are expressed through the art, craft, language, music, and varied cultural themes. The art of a society is influenced by the philosophical thought that governs its psyche. As a result, the comprehension of the same idea can differ in diverse philosophies like the Western and the Indian. The Western philosophy is based on a rational outlook and logic, and this directly or indirectly sustains and imparts a cultural identity to it. Their art and culture reflects this rational attitude. On the other hand, the Indian philosophy is based on traditional texts and scriptures, which are mystical and make the Eastern societies distinct from the West. Their growth and sustenance is based on the layers of beliefs that are entrenched since ages. The art and culture embodies the mysticism of varied beliefs and contexts. Consequently, one similar idea will be represented differently as an expression of art in diverse contexts. The art of a place will always reflect the cultural ethos of the region or context. Art will have a meaning if it relates to and is interpreted in a particular cultural background. This paper compares the Indian and Western art that is based on their respective understanding of the idea of "infinite." Through the examples, it exemplifies how a deeper understanding of cultural identity is important for a deeper understanding of art.

Keywords Philosophical thought • Logic • Mystical • Infinite • Western art • Indian art • Cultural identity

V. Sehgal (✉)
Dr. APJ Abdul Kalam Technical University, Lucknow, India
e-mail: sehgal_vandana@hotmail.com

F. Seta et al. (eds.), *Understanding Built Environment*, Springer Transactions in Civil and Environmental Engineering, DOI 10.1007/978-981-10-2138-1_21

21.1 Introduction

The "cultural identity" paradigm contains within it man's interaction with his context at economic, social, and environmental levels. Culture encompasses what people "have," that is, their material possessions. Culture is driven by what people "think," that is, their ideas, values, and attitudes. Culture is visible in what people "do," that is, how they behave (Andreatta and Ferraro 2009). The ideas and value system, over a period of time, get ingrained, and the thought process of the whole society is similar. "Cultural heritage connects people to a place that symbolizes the identity and values that provide a sense of belonging on personal and community levels, and the continuance of that heritage is what cultural sustainability is about" (Throsby 2003). Art in a society is driven by what people think and is a resultant of the ideas in their mind. These ideas are and have been influenced by the "deep structures," which are formed since time immemorial and are unique to a particular context. These ideas are portrayed in the philosophical texts of a particular society, and consequently, they form the basis of art and architecture as they create the context. The Western philosophical attitudes are different to the Indian way of thinking, which makes their people understand one similar idea differently. Therefore, one abstract idea can be understood in diverse sense in both the cultures. To understand a society culturally, it is important to understand these deeper attitudes.

21.2 Understanding of the Abstract Idea of "Infinite"

The abstract idea of "infinite" has different connotations in Western and Indian culture that is reflected in their art and culture and societal behavior. Their creations are derived from the deep-set attitudes and are reflected through them.

21.2.1 Indian Culture and "Infinite"

To understand the idea of infinite in Indian context, one needs to understand the background of Indian thought, which has religious overtones and is tinged with metaphysical connotations. This gives a holistic picture of the basis of the thought process that has taken centuries to formulate, which has stemmed from philosophical and metaphysical abstract concepts held in common across India and consciously or unconsciously applied in art.

In early *Upanishads*, which are the follow-up texts of the *Vedas*, the principal *Hindu* philosophical texts of Indian philosophy, the universe is referred to by the neuter term *Brahman*, which is an equivalent of an impersonal absolute that might also be called Oneness or Being. The expression "*atman* is *Brahman*" unequivocally identifies essential self with cosmos (Hamilton 2001). This concept has pervaded all traditional art, which in turn influences the contemporary trends, not so much as a stylistic idiom but as a deep structure that governs archetypes and molds thinking process.

The major distinguishing factor in creating and perceiving traditional Indian art and architecture is the involvement of the "inner self" – the *Tatatma.* The *bhav* (emotion), *vibhav* (practical sense), and *anubhav* (experience), all are to be synthesized (*sanyog*) in proper proportions (*matra*) to get *Rasa* (holistic experience). The Indian philosophy defines a piece of art as something that breathes and makes you respond through all the five senses to internally touch you. The preconceived abstract notions become measures that are intuitive and subjective. So the basic module rises above the normal experience and becomes more symbolic than realistic.

There are symbols that represent, manifest, and connote this connection between the self and the universe, which are part of the archetype in Indian art and architecture. One of them is the *mandala*, which represents the ideal world in a two-dimensional painted form or three-dimensional sculpture or built form. The traditional *mandala* is based on a religious philosophy or ritual. A *mandala* is organized around a central point, which symbolizes the energy center from where the existence is deemed to have imploded. It is also the symbol of the infinite and the absolute. The *mandala* (Fig. 21.1) presents a model of the macro- and microcosm in its visualization of the universe. Thus the *mandala* is a "cosmogram" which represents the essential schema of the entire universe (Jung 1977).

Graphically, the *mandalas* depict the whole universe symbolically. The emphasis of equating the 0 or the *bindu* with the infinite has been part of the Hindu philosophy. *Bindu* has been symbolic of a potent atom of energy that is about to implode.

Fig. 21.1 Mandala (http://www.geniusawakening.com/wp-content/uploads/2014/04/hindu-mandala.jpg accessed on 25.03.2015)

So, what we see is something tangible, which has the potential to convert into the intangible infinite. Symbolism has been and is the basis of Indian art and culture, which may be used as a tool for sustainable systems and implementing them.

21.2.2 *Western Culture and "Infinite"*

In the Western context, there has always been a rational leaning in philosophy, whose foundation was laid by the Greek philosophers. René Descartes, a French philosopher, who is associated with Cartesian geometry, took infinite to another level. He extended the range of infinity to the third dimension (Fig. 21.2). The rationality gave way to the scientific and mathematical thought process and in turn influenced the artists through the ages.

John Wallis, an English mathematician, while writing on the concept of conic sections, just threw in the symbol for infinity as ∞ and later reiterated it in his work on infinite series (Clegg 2003, p. 71). The diagrammatic depiction of infinity had been realized long back and was used in many areas like design and geometry as in

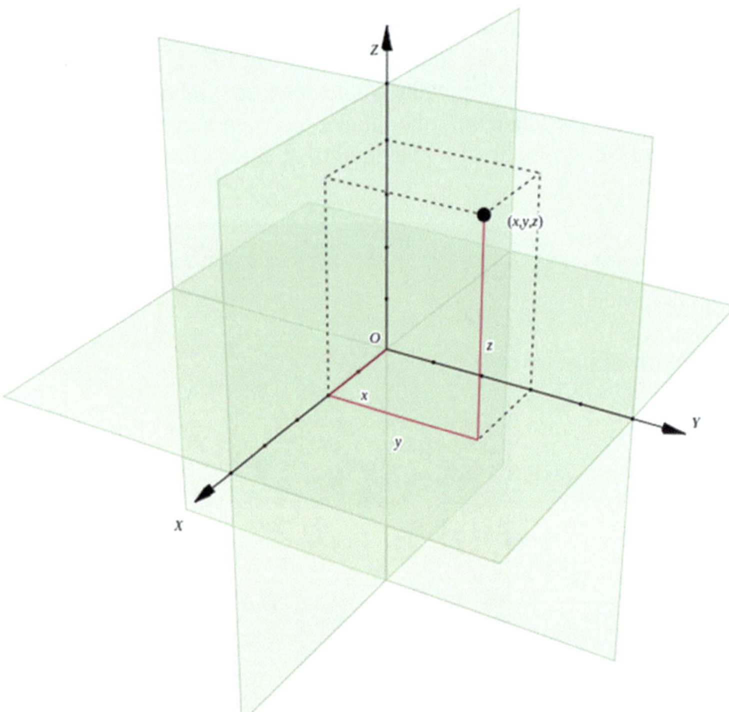

Fig. 21.2 Descartes model of 3-D Cartesian planes (http://simple.wikipedia.org/wiki/User:Jeffwang/Cartesian_coordinate_system accessed on 26.03.2015)

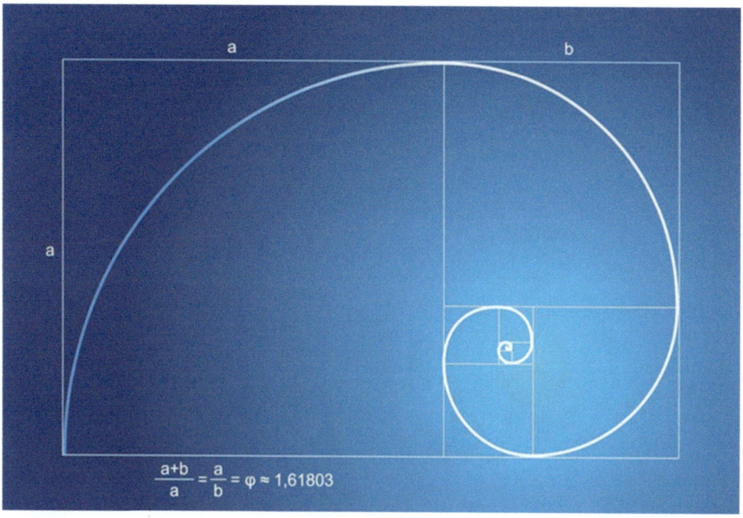

Fig. 21.3 Golden spiral (http://www.webdesignstuff.co.uk/mj248/2012/01/25/grids-in-webdesign-aesthetic-grid-a-grid-following-the-golden-ratio/ accessed on 26.03.2015)

the golden rectangle, the fractals, and the Von Koch curve. To construct a golden rectangle, a unit square is taken and a similar one is added to one of its sides; then a square with the sides equal to the sum of the previous two sides is added and this process can go on endlessly (Fig. 21.3). A spiral is constructed by joining the points in a golden rectangle that lie on the line that cuts it at one-third. Physically, it is impossible to continue for long, but as a process, it can continue forever to form a golden spiral.

A fractal is a rough or fragmented geometric shape that has a property called self-similarity, which means that it can be split into parts, each of which is approximately a reduced-size version of the whole. In mathematical terms, a diagrammatic depiction of exponents and logarithms is the fractals. The Van Koch curve is one such fractal. It is a construction of an equilateral triangle, whose sides are divided in three equal parts. The central one-third part of all the sides has equilateral triangles, and this phenomenon continues endlessly but does not go beyond the circle circumscribed around the initial triangle. This construction can go on infinitely and takes the shape of a snowflake (Fig. 21.4).

In art, the fundamental rules of "perspective" (Fig. 21.5), elaborated by Brunelleschi and Alberti, changed the way man perceived (Fig. 21.6). The fixed viewpoint of perspective in maps and paintings is elevated and distant, completely out of plastic or sensory reach. A conception of the finite space allowed the globe to be grasped as a finite totality.

Daniel Gran, an Austrian artist, shows us through his sketch (Fig. 21.7). A wall is never simply a wall, nor a ceiling, a ceiling. Each architectural element is extended beyond its functional duty as a shield from the hostile elements. The aesthetic component

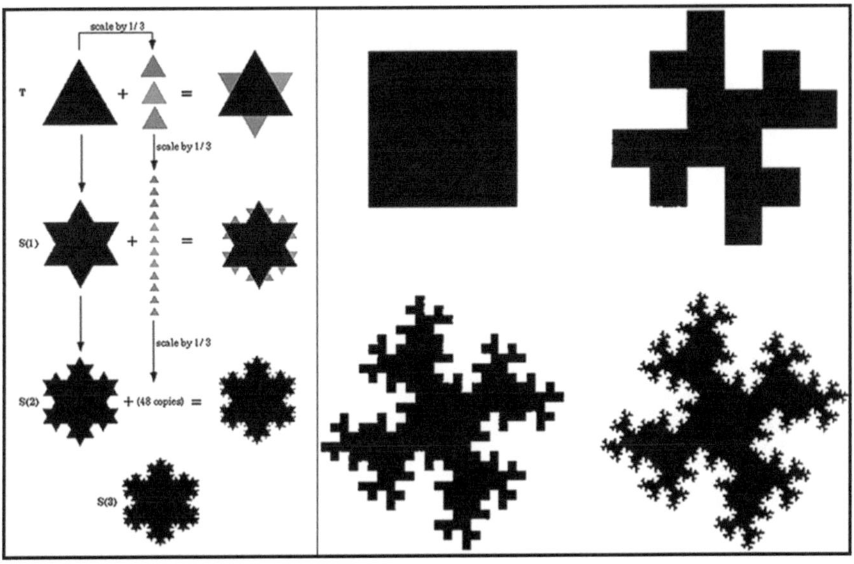

Fig. 21.4 Van Koch curve snowflake and a box fractal (http://t2.gstatic.com/images?q=tbn:ANd9GcRqxWGnInclkreL32fDFuGpmZP_lzWSRLHN0H4viiVAVz8Yk0&t=1&usg=__1zq-gRoUgTwFErxiSrg4-vaZ_04= accessed on 16.11.10)

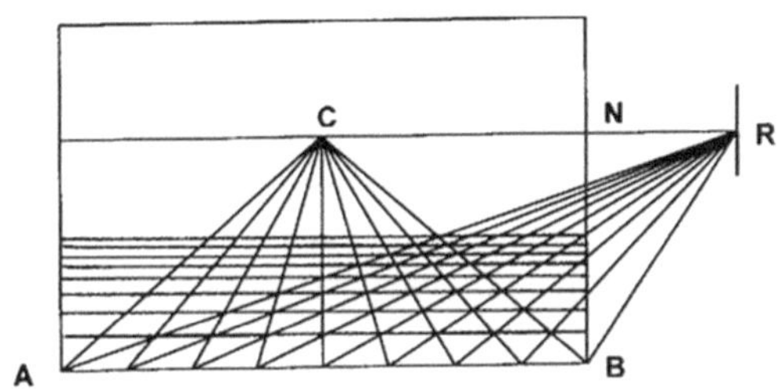

Fig. 21.5 Alberti's construction of perspective (http://www.math.nus.edu.sg/aslaksen/projects/perspective/alberti_files/image004.jpg accessed on 24.07.10))

of the object, its form, overtakes its function. A wall or a ceiling becomes a possible opening onto the reality, which it occludes:

> Contrast is the primary tool through which baroque art prompts a sensation of the infinite in the mind of the beholder. The infinite cannot of course be shown. It must be suggested or implied. What baroque art conveys is an impression, an illusion of infinite space, of movement into boundless depths, by suggesting the existence of what finally remains unseen. Contrast of light and dark, or chiaroscuro, gives space particular qualities. It accentuates the illusion of depth, giving the objects depicted a greater sense of mass and weight while simultaneously heightening their three-dimensionality, making them appear to jump out of the picture frame, or in the case of sculpture or decoration, out of the immediate space that "contains" them. (Norman 2001)

A Swiss sculptor, Max Bill, recreated the Mobius strip, which is an unending ribbon. "I created a single-sided object by searching for a solution of a hanging

Fig. 21.6 Raphael's "The School of Athens" depicting perspective (http://www.math.nus.edu.sg/aslaksen/projects/perspective/flagellation_files/image006.jpg accessed on 24.07.10)

Fig. 21.7 The Austrian National Library: Fresco painted by Daniel Gran in Baroque sensibility (http://jingalex.files.wordpress.com/2007/08/bibliotheck05.jpg accessed on 24.07.10)

Fig. 21.8 Max Bill's "Endless Ribbon" (http://www.bluffton.edu/~sullivanm/baltimore/baltimore1.html accessed on 07.10.09)

Fig. 21.9 Escher's "Treppenhaus" (http://www.mcescher.com/Gallery/gallery.htm accessed on 07.10.09)

sculpture, turning in the rising air. My search was neither scientific nor mathematical, but purely aesthetic…I named my sculpture *Endless Ribbon*" (Fig. 21.8).

He termed it as a "unilateral polyhedra" and, when he was compared with Mobius, said "What I could not find in Mobius' explanation is of primary importance to me: the philosophical aspects of these surfaces as symbols of infinity" (Maor 1987).

Maurits C. Escher was obsessed with graphics (Fig. 21.9) that showed no distinction between the solid and the void. More relevant are his sculptures (Fig. 21.10) where he distorted predictable forms to get unending surfaces.

The Western art is more logical and scientific in representing "infinite." This aspect of its rational psyche is important to understand the direction that is needed for its cultural sustainability.

Fig. 21.10 Escher's cube (http://www.bluffton.edu/~sullivanm/baltimore/baltimore1.html accessed on 07.10.09)

Fig. 21.11 Kundalini and bindu by S. H. Raza (http://www.paletteartgallery.com/upload/artsmall1504.jpg http://www.paletteartgallery.com/upload/artsmall2.jpg accessed on 05.08.10)

21.3 Contemporary Art

S. H. Raza, a contemporary Indian artist, paints with the traditional Indian sensibility. The point, the *bindu*, symbolizes for him the seed, the precursor to all life, or the epicenter of rhythmic reverberations. To express this concept, Raza has integrated a basic sense of geometry into his canvas, as is visible in his painting *Bindu* (Fig. 21.11). The infinity he connotes is the one that implodes inward and at the same time seems like an energy center. This is propagated by the Hindu ancient texts.

Fig. 21.12 Mondrian's "The Red, Yellow, and Blue" (http://en.wikipedia.org/wiki/Piet_Mondrian accessed on 23.01.11)

In the words of Ashok Vajpayee, eminent Indian poet, art critic, and writer:

> The finite space of the painted canvas uncurtains the immense infinitude that lies beyond conscious awareness. The viability of the painted space reflects the hidden energy… The connectivity with the infinite, with the hidden and the larger provides Raza's art with a dimension, which can only be called spiritual… It is a prayer which seeks 'dynamic equilibrium as expressive of the true nature of reality' as Mondrian would have had. (Vajpayee 2007)

Mondrian's paintings continued, in theory and perception, beyond the limits of the canvas. In his later works, the distinction between the figure and the background disappears, and the entire surface of the picture is organized as a single visual entity. He worked with composition of vertical and horizontal lines which intersect to form squares and rectangles in black, white, and primary colors (Fig. 21.12). Their rhythm is achieved by a grid pattern that seems to extend indefinitely in all directions giving an allusion of the infinite.

In one of his most stark and challenging paintings, *Fox Trot A* (Fig. 21.13), this is achieved with only three straight black lines on white, the vertical and horizontal cross near the right-hand bottom edge. This implies a similar crossing beyond the edge on the left, and the grid pattern thus formed could be indefinitely prolonged in all directions so that what we see is part of a larger whole, perhaps the infinite (Honour and Fleming 1984, p. 599). This effect is reinforced by the "arbitrary" lozenge shape of the canvas, which prevents the edges from engaging with the orthogonal pattern of lines. Schapiro says:

> In this art that seems so self-contained and disavows in theory all reference to the world outside the painting, we tend to complete the apparent forms as if they continued in a hidden surrounding field and were segments of an unbounded grid. It is hard to escape the suggestion that they extend in that virtual space outside. (Padovan 2002, p. 33)

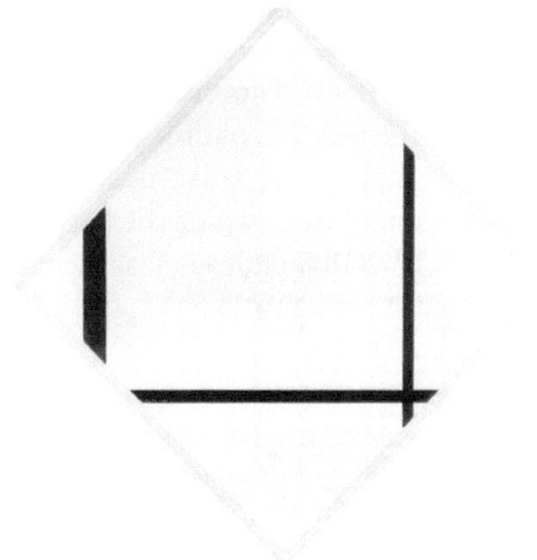

Fig. 21.13 Mondrian's "Fox Trot A" (http://www.artchive.com/artchive/M/mondrian/mondrian_fox_trot.jpg.html accessed on 12.03.10)

21.4 Comparison Between Western and Indian Art in Representing and Perception of "Infinite"

When "infinite" is represented in art, it is influenced by the contextual thought. The intention and perception of signifying an abstract idea like "infinite" is different in Indian and Western art because of cultural differences. Gilles Deleuze writes, "Our civilization is not one of the image but the civilization of the cliché." We often read images or texts on the level of the metaphor and perceive meaning as something fixed and stable, already existing. It is almost impossible to see an image as an image, without metaphor or preconstructed meanings. As a result, our context, philosophical leanings, and cultural background always color our reading or perception of art. Perceptual psychologists have studied the nature of cognitive structures that help perceive visual information and have found, after the analysis of common behavior patterns, that there is an existence of guiding structures of "expectations," "cognitive maps," or "deep structure." This implies that a particular "stereotyped" vision sees only those patterns, which its "stereotypes" have permitted it to anticipate. Visual codes that trigger the expectations are so firmly rooted in our seeing and perceiving that no crisis of representation arises (Salen 1993).

When we analyze the representation and perception of infinite in Western art and Indian art, the artist conceives them with a palimpsest of cultural background, and furthermore, the perceiver sees and comprehends it with his understanding. It is known that the paintings of Mondrian represent the "infinite" through the illusion of the never-ending grid and the canvass appearing to be a small part of an endless painting. In his painting, *Fox Trot A*, this is achieved with only three straight black lines on white, the vertical and horizontal cross near the right-hand bottom edge. This work, like most of his paintings, has the incomplete grid that one wants to complete and in the process, inadvertently, brings in the surrounding space and beyond into the realm of perception. Mondrian's space makes one think of the Cartesian transparency and at the same time alludes to the philosophical lineage and abstraction of Plato. This never-ending infinite space is part of the rational and scientific way the Western philosophy looks at infinity. On the other hand, H. M. Raza, a contemporary Indian artist, uses the "*bindu*," which is a sparkling, infinitesimal dot, or the spark, from which the universe unfurls and into which it curls back. The Hindu religious thought says that the energy, time, and space, and perhaps the first light followed by the first sound, came from a *bindu*. The onlooker is sucked into the never-ending vortex of Raza's *bindu* or the *kundalini* series. These paintings of Raza do not depict the immediate reality and issues but explores and enacts the more abiding aspects of life and existence. It goes to the origins and rearticulates the primordial insights. (Vajpayee 2007) To comprehend these series by Raza, one has to be familiar with the Hindu philosophy of internalizing the energy and the allusion of the infinite in the geometrically contained circle. Both the artists, Mondrian and Raza, represent the two diverse comprehensions of the "infinite" in the Western and the Indian thought – the explosion and the implosion, respectively – and they are a reflection of their cultural ethos and resultant philosophy. To appreciate and perceive them, one has to understand the theoretical background and the philosophical thought that informs because the artists have lived their philosophy and culture that inspires their work.

21.5 Conclusions

The paper reflects on the diverse understanding of the abstract idea of infinite in diverse cultures like the Western and the Indian. Subsequently, it elucidates the manifestation of this idea of infinite in their respective art. The fact is that the same idea of "infinite" is portrayed by the Western and Indian artistes differently. The basic difference in western and Indian philosophy has been elucidiated through their respective contemporary art and sculpture, where the difference between them is the western logical attitude versus the mythologically steeped Indian thought. This is elucidated through the works of Mondrian, whose paintings allude to the "infinite" through his rational composition of unending grid, where the canvas is a small part of it. On the other hand, in the Indian artist, Raza's paintings represent the "infinite" through the symbolic *bindu*. This showcases the fact that each culture is different because of the difference in basic beliefs, value systems, and traditions and the artists' language of expression consciously or unconsciously employs that layer. Understanding of art requires a deeper comprehension and sensitivity to this cultural layer. What is sacred or important in one culture might not be a priority for another. It was established that one perspective cannot hold good for perceiving art from diverse cultures.

References

Andreatta S, Ferraro G (2009) The concept of culture – elements of culture: an applied perspective. Wadsworth Publishing, Belmont

Clegg B (2003) A brief history of infinity: the quest to think the unthinkable. Constable and Robinson, London

Hamilton S (2001) Indian philosophy: a very short introduction. Oxford University Press, New York

Honour H, Fleming J (1984) The visual arts: a history. Princeton Hall, New York

Jung, CG (1977) *Collected works of C.G. Jung*, vol.18, Configurations of the Unconscious. Princeton University Press, Princeton

Maor E (1987) To infinity and beyond. Princeton University Press, Birkhauser/Boston

Norman LF (2001) The theatrical baroque: European plays, painting and poetry, 1575–1725 (online) Available at: http://fathom.lib.uchicago.edu/2/10701023/. Accessed on 25.03.2015

Padovan R (2002) Towards universality: Le Corbusier, Mies+De Stijl. Routledge, London/New York

Salen, K (1993) Speaking the text: the resonance of syntactic difference in text interpretation, Virginia Commonwealth University

Throsby D (2003) In: Towse R (ed) A handbook of cultural economics. Edward Elgar Publishing, Northampton

Vajpayee A (2007) Reflections. In: Raza, selected works. Art Alive Gallery and N.G.M.A, New Delhi

Chapter 22
Local Residents' Perception of Social and Economic Impacts of Urban Riverfront Development: Case of Sabarmati Riverfront Development Project

Smriti Mishra, Jaydip Barman, and Shashi Kant Pandey

Abstract Waterfront development projects have an all-round impact on cities. Many scholars have acknowledged this; however their research has not considered the perception of local residents, about such developments and its impacts. This paper therefore addresses this imbalance in the literature by analysing a survey of local residents' perception of the social and economic impacts of such developments by considering India's first riverfront development project, the Sabarmati Riverfront Development Project (SRFD), Ahmedabad. The research findings examine the structural relationship between positive and negative dimensions of social and economic impacts with the total impact dimensions of urban riverfront development as perceived by the respondents. Among the twelve perceived economic impact and fourteen perceived social impact exogenous constructs, only eight and six constructs respectively are identified as strong components contributing to the total impact perception of urban riverfront development. The rest of the positive social impact dimensions are not perceived in a significant manner. Approaches should be considered to nullify the negative perceptions and to intensify the positive construct by involving the community or by considering public participation in the whole planning process of urban riverfront development. The study also highlights that though 74 % of the respondents support mixed-use development for the riverfront, there is no significant relationship between this preference and perception of overall impact of urban riverfront development. The conclusive remarks highlight the need to consider residents' perception at the very inception stage to ensure their support and participation in the planning and development of such projects.

S. Mishra (✉) • S.K. Pandey
Department of Architecture, Birla Institute of Technology Mesra, Ranchi, Jharkhand, India
e-mail: smriti151@yahoo.com

J. Barman
Department of Architecture and Regional Planning, Indian Institute of Technology, Kharagpur, West Bengal, India

F. Seta et al. (eds.), *Understanding Built Environment*, Springer Transactions in Civil and Environmental Engineering, DOI 10.1007/978-981-10-2138-1_22

Keywords Urban waterfront development • Riverfront • Economic impact • Social impact • Resident perception • SRFD

22.1 Introduction

Waterfront development is a prominent worldwide phenomenon in the "postmodern" urban context (Norcliffe et al. 1996) and an element of the late-twentieth-century urban development initiatives. It is a story of growth, decline, regeneration and the new found success which has been an issue of wide concern and extensive discussion since the 1970s. In the pre-industrial and the industrial era, urban waterfront areas were the focal point of various water-dependent and water-independent activities, which were normally and intimately related both in functional and spatial terms (Hoyle and Pinder 1992a).

This once well-established symbiotic relationship persisted until the mid-twentieth century (Hoyle 1997). It got severed in the post-industrial era after World War II, due to transitions in three aspects: expansion of city size, reforms in transportation technologies and changes in industry. Thus, from being one of the major nodes of the city, the waterfront became a derelict area facing severe pollution, low accessibility and a poor image, leaving the dilapidated warehouses, factories and port facilities to further decay. Cities facing changing relationships and trying to respond to the phenomenon of post-modernism recognised waterfronts for their socio-economic development potential. These developments were widely perceived as potential economic supports that have the capacity to redefine the identity of the city; provide opportunities that may improve the quality of life; provide employment opportunities; offer economic diversity, tax revenues and business opportunities for festivals, restaurants, natural and cultural attractions and outdoor recreation; and improve the environmental condition and enhance its competitiveness and vitality. Cities today, with any form of water, are thus making an effort to redevelop and reconnect to their waterfront areas (Hoyle 2000). Thus cities started to transform their once industrial or otherwise waterfront areas into vibrant zones of leisure, commerce and housing (Bruttomesso 1993; Breen and Rigby 1994, 1996; Craig-Smith and Fagence 1995; Malone 1996). However, there were also serious concerns that it can have negative impacts on the quality of life in the form of crowding, traffic and parking problems, increased crime, increased land prices in surrounding areas, increased cost of living, conflict between tourists and residents and alteration of hosts' lifestyles. Massive urban regeneration efforts were first undertaken in cities throughout the USA. The earliest efforts started in San Francisco in the 1950s, but the real potential of waterfront regeneration gained recognition with the efforts done in Boston and Baltimore in the USA.

However, scholars, researchers and waterfront communities have often raised concerns that such development does not always include the interest and the desire of the local residents and are not socially inclusive. Breen and Rigby (1996) highlighted that urban waterfront development have not well-served poor people as the space on the waterfront generally gets gentrified. Referring to cases of Tiger Bay in Cardiff, the Isle of Dogs in London and along the river in Singapore, they have mentioned that communities were ignored, relocated or, worse, displaced as new and more wealthy populations moved in the waterfront development area. Regarding the architectural and spatial form of urban waterfronts, the designs follow global references, preferring an international style more than local references (Martire 2009); they are homogenous and standardised waterfronts, with little or, worse, without any kind of local identity. It is more about production of public space on the urban waterfront. It appears that, on one hand, the transformation of urban waterfronts is intended to reconnect people to their waterfront, but on the other hand, this intention also represses the desire of certain group of people especially, the local community, on their own waterfront.

The tale of such areas in the developing countries is even more pathetic. They are under huge pressure, in maintaining a balance between the socio-economic and environmental issues, often failing miserably. Considering the Indian subcontinent, rivers or other waterfront areas have always had great sociocultural and religious significance. However, the most disturbing fact is that most of the country's waterfront areas present a dismal scenario. Indian rivers, for example, are more like canals today, taking all domestic and industrial waste of the settlement. These areas lie neglected and are often characterised by unplanned development and incompatible land uses. The insufficient infrastructure and inefficient planning policies along these areas have further aggravated the situation, thereby transforming most of the area into blighted areas. So, how do people, especially the local residents, react on these issues? How does the social dimension of urban waterfronts really work as public spaces in these conditions? What kind of social activities and interactions and benefits are desired by the local residents on the urban waterfronts? How do people use and think of public space on urban waterfront? How do people perceive the economic and social benefits of riverfront development and the kind of development type they want to have on their riverfront. Thus, the need to answer such questions leads to the construct of the aim of this research.

Therefore, with an objective to identify the perception of local residents regarding the social and economic impact dimensions of urban riverfront development, this paper presents findings of an empirical study of a survey conducted in Ahmedabad, India, which is the first city in the country to have a riverfront development project. The value of this case study lies in the fact that Sabarmati Riverfront Development is a first of such effort considered for a city on a river in India. Conducting research at this point of time when residents have had an experience with the ongoing riverfront development project will give an important insight about the expectations of the local residents of such projects. The understanding developed from this research will further help in reinforcing and enhancing such future efforts when other Indian cities are also trying to work out such options.

The tenet stipulated in this study is that perceived positive and negative dimensions of economic and social impact of riverfront development will influence the perception of each other and the perceived total impact of riverfront development in varying degrees and in different directions. Also each impact factor will have

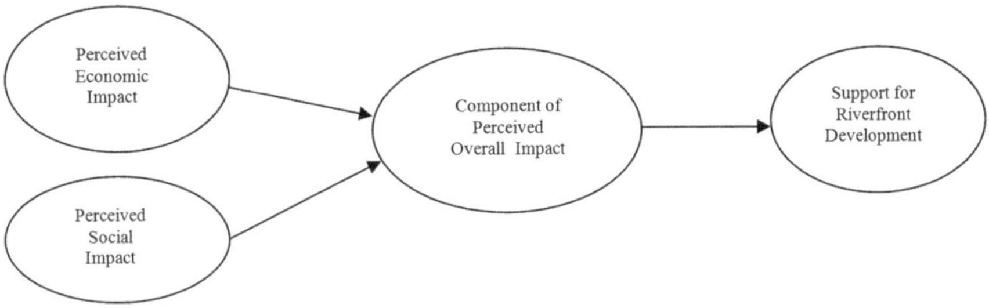

Fig. 22.1 The proposed theoretical model

varying effects on local residents' support for a certain type of urban riverfront development model which are mediated by the perceived total impact of riverfront development. This research therefore proposes to examine the structural relationship (refer Fig. 22.1) between the perceived dimensions of social and economic impacts and the perceived total impact dimensions of urban riverfront development and support for a certain type of riverfront development by the respondents. The research hopes in contributing by identifying ways in which it will improve and ensure increased public participation and public support in the whole process of planning and development of such projects.

22.2 Research Methodology

This section discusses the research methodology adopted for this paper. The whole study is done in two parts: first is the selection of case study area of urban riverfront development. Since Sabarmati Riverfront Development Project, Ahmedabad, is the only urban riverfront development project which has been fully initiated in India, therefore the study has selected it as a case study. The second part of the section focuses on the data collection and analysis techniques adopted for the study.

22.2.1 Case Study: Sabarmati Riverfront Development Project, Ahmedabad

The city of Ahmedabad, having a population of more than 5.8 million and an extended population of 6.3 million, is the largest city in the state of Gujarat located on the banks of the river Sabarmati, which has served as Gujarat's lifeline for ages. It was long felt that the Sabarmati River and the appropriate development of its riverfront can turn the river into a major asset. The Sabarmati Riverfront Development Project was conceived in 1998, as a pioneering multi-benefit multidimensional project with significant urban transformation potential. It however gained momentum only in 2001 after the initiation of then chief minister and now prime minister Mr. Narendra Modi. Construction of the project finally started in 2005. The project plans to upgrade 18 precincts, in the heart of Ahmedabad, leading to the city's revitalisation and hence its future growth. With an ambition to reconnect the city with the river, 202.79 ha of riverbed land on both the banks has been reclaimed to create a public riverfront on both the banks with a two-level, continuous promenade on both sides of the river. The multidimensional riverfront project aims to provide city-level social infrastructure by rehabilitation and resettlement of riverbed dwellers and activities, creation of parks and public spaces and provision of sociocultural amenities for the city; provide for environmental improvement by considering reduction in erosion and flood control features to safeguard the city, sewage diversion to clean the river and water retention and recharge of the river bed; and include sustainable development aspects as an objective for the project and city at large by aiming at generation of resources and revitalisation of its neighbourhoods. The project cost in 1997 was worked out to be Indian rupees (INR) 3610 million; however constant changes in the plan and other hindrances escalated it to an estimated INR 12,000 million today.

22.2.2 Data Collection and Analysis

A questionnaire survey was conducted in Ahmedabad City to explore the perception of the local residents about the positive and negative overall social and economic impact dimensions of the urban riverfront development project for the city. A large sample is always preferred for structural modelling, and in this context it is suggested that the model is acceptable if a minimum ratio of at least five respondents for each estimated parameter or a more appropriate ratio of ten respondents per parameter is obtained; however many other factors such as model misspecification, model size, estimation procedure and normality issues impact sample size requirements (Hair et al. 2008). Therefore in order to meet the sample size for the study, a total of 1000 survey questionnaires were distributed to the target population, and some were completed by face-to-face interview. The survey was conducted in between March 2014 and August 2014. Since the questionnaire included 36 research parameters, the received response of 474 completely filled questionnaires met the requirement. The study therefore proceeded with the analysis of data by grouping the absolute age of the different respondents into these age bands: 18–24 years old, 25–34 years old, 35–44 years old, 45–54 years old, 55–64 years old, 65–74 years old and more than 75 years old.

The entire questionnaire was designed into four sections: first of introduction, second to know the respondent's socio-economic and demographic profile, third to gauge the local residents' perception of the overall, social and economic impact of urban riverfront development and the fourth to know the kind of development they would support for their riverfront. The questionnaire included two questions to assess the perception of local residents of Ahmedabad regarding the overall impact

Table 22.1 Perception dimensions of the endogenous construct overall impact of urban riverfront development

S. no.	Item code	Description of items
1.	TI01	Perception of the community about overall impact of riverfront development
2.	TI02	Perception of the community if the benefits of riverfront development are greater than the costs of its development

Table 22.2 Perception dimensions of the exogenous construct social impact of urban riverfront development

S. no.	Item code	Description of items
1.	SI01	Riverfront development will improve the quality of life of urban residents
2.	SI02	Riverfront development will bring benefits for all city residents through generation of wealth, jobs, an improved built environment and physical infrastructure
3.	SI03	Riverfront development will create new public spaces and new places to visit, admire and enjoy
4.	SI04	Riverfront development will increase access to valued cultural and natural amenities; water, a natural element, will become more accessible
5.	SI05	Riverfront development will create and promote a more positive image of the city
6.	SI06	Riverfront development will displace low-income and ethnically different populations from the riverfront area
7.	SI07	Riverfront development is a misuse/waste of taxpayers' money
8.	SI08	Riverfront development will attract visitors/tourists to the city
9.	SI09	Riverfront development will increase the imageability of the city with its iconic identity
10.	SI10	Riverfront development will make the riverfront area less accessible to less privileged people and will lead to social segregation
11.	SI11	Riverfront development will make the riverfront area crowded beyond its natural threshold/bearing capacity
12.	SI12	Riverfront development will provide with more of community space for social/community interaction
13.	SI13	Riverfront development will influence the safety of the user, and there will be more occurrences of criminal activities
14.	SI14	Riverfront development will shift the focus from water-dependent uses to non-water-dependent uses

of urban riverfront development (refer Table 22.1). It also includes 14 questions to assess the perception of the local residents about the social impacts and 12 questions to assess the economic impacts of such development (refer Tables 22.2 and 22.3, respectively). The respondents had to rate each impact question on a 5-point Likert scale, where 1 represented strongly agree and 5 represented strongly disagree. The residents were also asked to rate the kind of riverfront development which they would like to support for their city (refer Table 22.4); 1 represented strongly oppose and five represented strongly support for the suggested type of development.

Table 22.3 Perception dimensions of the exogenous construct economic impact of urban riverfront development

S. no.	Item code	Description of items
1.	EI01	Riverfront development will increase the city's value in terms of economic competitiveness compared to other cities in the region
2.	EI02	Riverfront development will create more jobs for your community
3.	EI03	Riverfront development will attract more investment to your community
4.	EI04	Riverfront development will lead to more spending in your community
5.	EI05	Riverfront development will give more economic benefits to local people and business establishments
6.	EI06	Riverfront development will increase the value of property in its adjacent area
7.	EI07	Riverfront development will boost the housing price in its adjacent area
8.	EI08	Costs of developing riverfront facilities are too much of a burden on government/public sector agencies
9.	EI09	Riverfront development will attract a lot of public and private investment
10.	EI10	Riverfront development will increase real estate tax revenues
11.	EI11	Riverfront development will offer substantial premiums to developers, landowners and local government
12.	EI12	Riverfront development will concentrate development and investment activity and economic regeneration only in few places along the riverfront within the city

Table 22.4 Types of riverfront development supported or opposed by local residents

S. no.	Item code	Description of items
1.	DT1	No development at all
2.	DT2	Only commercial riverfront [retail commercial, shopping precincts, shopping plaza, street shopping]
3.	DT3	Only cultural, educational and environmental riverfront
4.	DT4	Only historic riverfront [retention and enhancement of historical elements and cultural heritage of the place]
5.	DT5	Only recreational riverfront [parks, gardens, picnic areas, walking, cycling and water-related activities including boating, fishing]
6.	DT6	Only working riverfront [water-dependent uses and activities – water-based transport, industries, commercial fishing, dhobi ghat]
7.	DT7	Only residential riverfront [single unit duplex, individual plotted, multi-storey]
8.	DT8	Mixed-use riverfronts [residential, commercial, recreational, industrial and working]

The exogenous constructs of perceived total impact and endogenous constructs of perceived social impact and economic impact factors were analysed on the basis of descriptive statistics like mean, standard deviation, skewness and kurtosis. Skewness and kurtosis is considered to test the normality of the variables. Normal probability plot compares the cumulative distribution of actual data, values with normal distribution which forms a straight diagonal line. Bentler (2005) as cited in Byrne (2010) suggested that the critical ratio which is a normalised estimate of multivariate kurtosis, with values more than 5, is indicative of data that are non-normally distributed. As a rule of thumb, Byrne suggested that variables can be considered as moderately non-normal if they indicate skewness values ranging from 2.00 to 3.00 and kurtosis values from 7.00 to 21.00, and extreme non-normality is defined by skewness values greater than 3.00 and kurtosis values greater than 21.00. Since the captured data is qualitative in nature, therefore mean and standard deviation are used for gauging the perception of the respondents only at an indicative level.

Further, the complete covariance matrix for all the exogenous and endogenous constructs was expressed in a symmetry matrix. KMO sample adequacy and Bartlett test of sphericity were done. Extraction method was done on the basis of principal component analysis of eigenvalue equal to 1. Varimax rotation method was applied with Kaiser normalisation. List-wise deletion was done based on all variables in the procedure. Further reliability statistics was done by Cronbach's alpha. Cronbach's alpha based on standardised items was found. Finally, ANOVA was conducted with Friedman's test and Tukey's test for non-additivity. All analysis was undertaken using SPSS version 17 software.

22.3 Results and Research Findings

Entire findings can be divided broadly into two sections: socio-economic and demographic characteristics of the respondents and perception regarding the impact of Sabarmati Riverfront Development Project.

22.3.1 Socio-economic and Demographic Characteristics of the Respondents

This section discusses the demographic and socio-economic characteristics of the respondents. Of the total 474 respondents, 71.7 % were males and 28.3 % were females. The age of respondents ranged from 18 to 75 years and above. The highest number of respondents involved in the study was of the ages 25–34 years old (33.4 %), and the second highest were in the 18–24 years old category (27 %), followed by 35–44 years old category (18.5 %) and 45–54 years old (11.3 %). The distance of residence of the respondents was also enquired. Three hundred twenty-two respondents stayed within a distance of 0–5 km from the riverfront, 97 from 6 to 10 km, 52 at a distance from 11 to 20 km and 3 who stayed at a distance of more than 20 km. Regarding family size of the respondents, 220 (46.4 %) of the respondents belonged to a family size of four persons; 175 (36.9 %) respondents belonged to a family size of more than four persons; 63 (13.3 %) respondents belonged to a family size of three persons; and the rest 16 (3.4 %) persons belonged to a family size of two persons. Survey shows 119 respondents (25.1 %) had higher secondary education; 109 respondents (23 %) were graduates; 99 respondents (20.9 %) were undergraduates; 42 respondents (8.9 %) were professionals; 28 respondents (5.9 %) were of technical background; 21 respondents (4.4 %) had attained education till class nine; 20 respondents (4.2 %) did not have formal education but could read and write; 15 respondents (3.2 %) had attained education still standard fifth and another 15 respondents (3.2 %) had other educational background other than listed; and 6 respondents (1.3 %) were illiterate. Regarding occupation of the respondents, 118 (24.9 %) were in private service; 85 (17.9 %) were students; 71 (15 %) were self-employed; 55 (11.6 %) were in government service; 43 (9.1 %) respondents were females who were housewives; 41 (8.6 %) are associated with petty trade; 17 (3.6 %) respondents were skilled workers; 17 (3.6 %) were retired people; 9 (1.9 %) were farmers, 7 (1.5 %) were unemployed; 6 (1.3 %) were engaged in different other occupation; and 5 (1.1 %) were wage labourers. As far as income is concerned, 116 (24.5 %) of the respondents had monthly income in the range of INR 5000–10,000; 176 (37.1 %) in the range of INR 10,001–20,000; and 109 (23 %) in the range of INR 20,001–50,000, and 73 (15.4 %) had monthly income more than INR 50,001.

22.3.2 *Respondents' Perception Regarding the Impact of Sabarmati Riverfront Development Project*

This section deals with the statistical analysis of the endogenous constructs, that is, the perception of the local residents of Ahmedabad regarding the overall impact of the Sabarmati Riverfront Development Project on the city, and the two components of exogenous constructs, that is, the economic impacts and social impacts of riverfront development and choice of development on the riverfront.

22.3.3 *Respondents' Perception Regarding the Overall Impact Dimension of Sabarmati Riverfront Development Project*

The descriptive statistics of two endogenous constructs (refer Table 22.5) show that TI01 has high skewness and kurtosis than TI02. In terms of positive skewness, the probability density function of both endogenous constructs has a long tail towards the right. Again kurtosis confirms that the constructs have sharper peak and flatter tail. However the perception of the community about overall impact of riverfront development (TI01) has comparatively longer and flatter tail to the right than the comparative perception of benefit cost perceived by the community (TI02). This kurtosis in both the constructs supports leptokurtic or Laplace type of distribution.

Table 22.5 Descriptive statistics of perceived overall impact constructs

Variable	Mean	SD	Skewness	Kurtosis
TI01	1.418	0.74	2.388	6.908
TI02	2.207	0.738	1.9	4.404

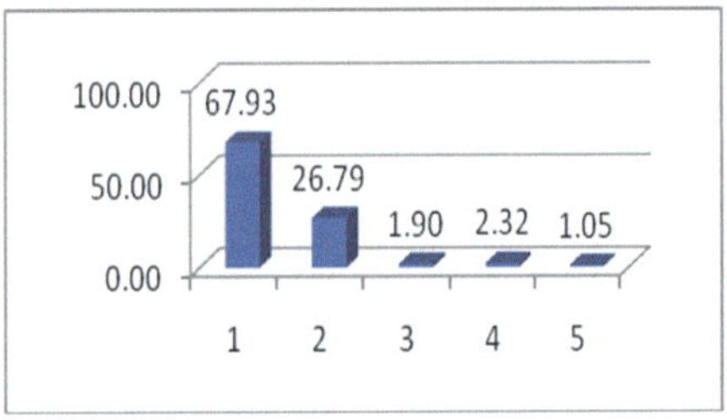

Fig. 22.2 Graph showing perception of the respondents regarding the construct overall impact of riverfront development (TI01)

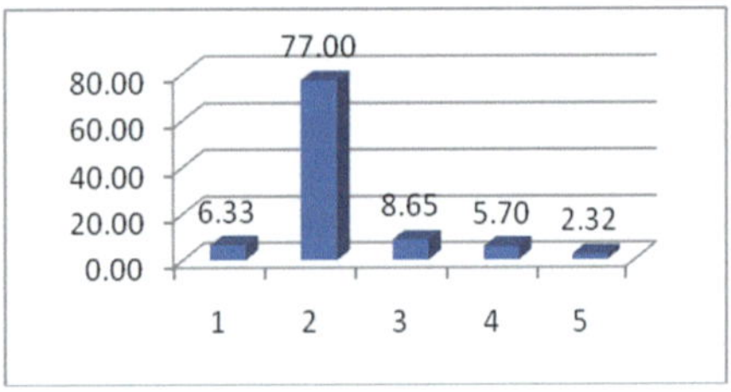

Fig. 22.3 Graph showing perception of the respondents regarding the construct benefits of riverfront development are greater than the costs of its development (TI02)

The normality test shows that the construct TI02 is normal, whereas TI01 is moderately non-normal (Figs. 22.2 and 22.3).

22.3.4 *Respondents' Perception Regarding Positive and Negative Social Impact Dimensions of Sabarmati Riverfront Development Project*

The descriptive statistic of exogenous construct perceived social impact (refer Table 22.6) shows that standard deviation is high for the construct that riverfront development will displace low-income and ethnically different populations from the riverfront area (SI06) followed by the construct that riverfront development is a waste or misuse of the taxpayers' money (SI07), that riverfront development will create and promote a more positive image of the city (SI05), and that riverfront development will make the riverfront area less accessible to less privileged people leading to social segregation (SI10). The normality test also confirms that all social impact constructs have a normal distribution except the construct SI06 which has an extreme non-normal distribution. The entire exogenous construct has a probability density function

Table 22.6 Descriptive statistics of perceived social impact constructs

Variable	Mean	SD	Skewness	Kurtosis	Rank
SI01	2.015	0.784	1.427	3.205	1
SI02	2.213	0.881	1.113	1.071	3
SI03	2.243	0.992	0.766	−0.048	5
SI04	2.399	0.992	0.869	0.08	7
SI05	2.285	1.055	0.875	0.06	6
SI06	2.447	1.357	6.592	90.043	9
SI07	2.876	1.228	0.225	−1.16	14
SI08	2.143	0.965	1.015	0.612	2
SI09	2.241	0.953	0.963	0.51	4
SI10	2.569	1.104	0.571	−0.656	13
SI11	2.466	0.942	0.654	−0.475	10
SI12	2.399	0.932	0.791	−0.012	7
SI13	2.528	0.967	0.726	−0.402	12
SI14	2.519	0.974	0.657	−0.522	11

long tail to the right, especially very high and long about the displacement matter. Here kurtosis confirms that the constructs about quality of life of urban residents and displacement of low-income and ethnically different populations from the riverfront area have high leptokurtic distribution, while the constructs about riverfront development bringing benefit for all city residents through generation of wealth, job and improved built environment and physical infrastructure have low leptokurtic distribution, but the rest of the constructs are either mesokurtic or platykurtic.

Careful observation of all the graphs of social impacts suggests that they can be grouped in three categories. The social construct that riverfront development will improve the overall quality of life of the urban residents has a single bell with 67 % of the respondents agreeing to this. The social constructs that riverfront development is a misuse/waste of taxpayers' money and that the riverfront area will be less accessible to less privileged people and will lead to social segregation are an example of significant double bells. A significant number of respondents, i.e. 26–38 %, disagree or strongly disagree with these constructs. The rest of the components follow the third category of double bell, but a very insignificant number disagree to the construct.

22.3.5 *Respondents' Perception Regarding Positive and Negative Economic Impact Dimensions of Sabarmati Riverfront Development Project*

The descriptive statistic of the exogenous construct of perceived economic impact (refer Table 22.7) shows that standard deviation of the construct that cost of developing riverfront facilities are too much (EI08) is high, followed by the constructs that riverfront development will give more economic benefits to local people and

Table 22.7 Descriptive statistics of perceived economic impact constructs

Variable	Mean	SD	Skewness	Kurtosis	Rank
EI01	1.89	0.808	1.281	2.568	1
EI02	2.12	0.778	1.250	2.263	3
EI03	2.13	0.946	0.945	0.713	4
EI04	2.26	0.985	2.148	12.823	6
EI05	2.26	0.997	0.751	–0.168	7
EI06	2.14	0.953	0.818	0.083	5
EI07	2.06	0.949	0.983	0.558	2
EI08	2.56	1.061	0.505	–0.667	12
EI09	2.28	0.991	0.891	0.280	8
EI10	2.35	0.984	0.743	–0.137	10
EI11	2.29	0.938	0.789	0.196	9
EI12	2.45	0.974	0.768	–0.201	11

business establishments (EI05), that it will attract public and private establishments (EI09) and that it will concentrate development and investment activity and economic regeneration only in few places along the riverfront development (EI12). Careful observations pertaining to the different perceived economic impact constructs suggest that majority of them follow double-bell-shaped distribution, with more than 50 % of the respondents agreeing with the construct on a 5-point Likert scale. The constructs, that the riverfront development will increase the city's value in economic competitiveness (EI01), will create more jobs for the community (EI02), will attract more investment (EI03), will lead to more spending (EI04) and will offer substantial premiums to developers, landowners and local government (EI11), all follow a single-bell distribution with the respondents mostly agreeing and strongly agreeing with the constructs. The normality test for the economic impact constructs shows that all the constructs have a normal distribution except the construct that riverfront development will lead to more spending (EI04), which has moderately non-normal distribution.

22.3.6 *Scree Plot, Component Matrix and Inter-item Covariance Matrix of Perceived Socio-economic Constructs*

KMO sample adequacy was done which was found to be 0.858. Bartlett test of sphericity was also conducted where chi-square value approximately was 3.117E3 with 378 degrees of freedom. The extraction method was done on the basis of principal component analysis of eigenvalue equal to 1. These extracted eight components (refer Fig. 22.4 and Table 22.8) which were able to explain cumulatively 59.286 % of the variance, 20.145 % from component 1 and 3.778 % from

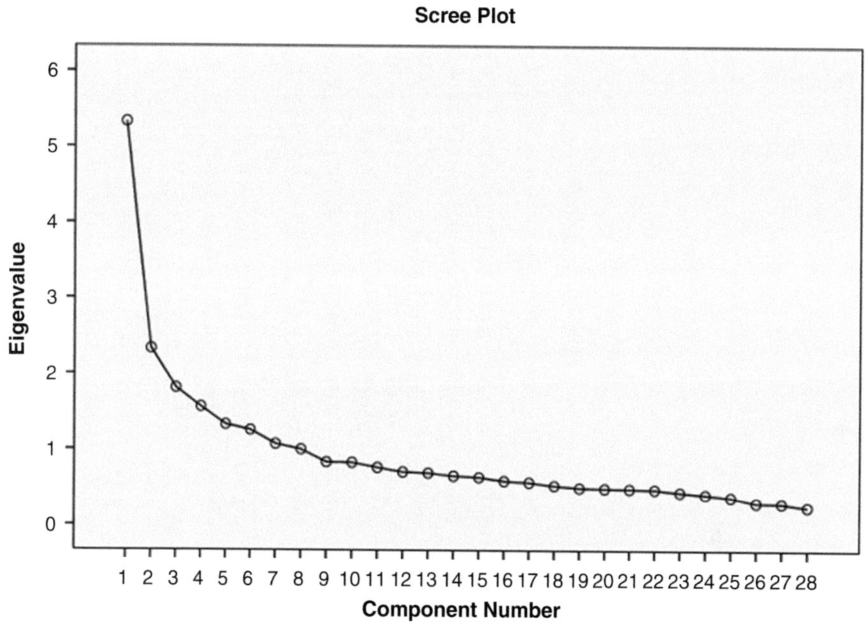

Fig. 22.4 Scree plot of perceived socio-economic impact constructs

component 8. Again varimax rotation method was applied with Kaiser normalisation. List-wise deletion was done based on all variables in the procedure.

Further reliability statistics was done by Cronbach's alpha, which was found to be 0.844, and Cronbach's alpha based on standardised items was found as 0.853 for 28 items. Item statistics (refer Table 22.9) show that the constructs have inter-item variance as low as 0.009, and the ANOVA table with Friedman's test and Tukey's test for non-additivity suggests low significance level, i.e. 0.104 (refer Table 22.10).

22.3.7 Respondents' Perception Regarding Type of Development They Would Support

Regarding respondents' perception regarding type of riverfront development that should come up on the riverfront or they would like to support, the first choice of no development on the riverfront and the single-use riverfront development choice constructs (like only commercial, cultural, educational, historic, recreational or residential) have a long-tail probability density function towards the right; however the development of mixed-use construct shows long tail to the left. In terms of kurtosis, sharper peak and flatter tails are found in some constructs like no development and only residential riverfront development (like logistic distribution). However, riverfront development choice constructs like development of only commercial or

Table 22.8 Component matrix of perceived socio-economic impact constructs

	Raw								Rescaled							
	Component[a]								Component							
	1	2	3	4	5	6	7	8	1	2	3	4	5	6	7	8
EI03	0.573	−0.090	0.024	−0.088	−0.011	−0.294	0.112	−0.103	0.605	−0.096	0.026	−0.093	−0.011	−0.311	0.118	−0.109
EI11	0.556	−0.068	0.159	−0.194	−0.051	−0.009	−0.169	−0.079	0.593	−0.072	0.169	−0.206	−0.054	−0.009	−0.181	−0.084
SI05	0.611	−0.160	−0.009	0.275	−0.359	0.215	0.237	0.183	0.579	−0.152	−0.008	0.261	−0.341	0.204	0.225	0.174
EI05	0.571	−0.030	0.163	−0.167	−0.116	−0.302	0.215	−0.163	0.573	−0.030	0.163	−0.167	−0.116	−0.303	0.216	−0.164
EI01	0.457	−0.199	0.151	−0.012	0.096	−0.227	0.121	−0.030	0.565	−0.246	0.187	−0.015	0.119	−0.281	0.149	−0.037
SI02	0.480	−0.103	−0.078	−0.028	0.040	−0.056	0.147	0.348	0.540	−0.116	−0.088	−0.031	0.045	−0.063	0.165	0.391
EI09	0.531	−0.177	0.163	−0.287	−0.253	−0.077	0.009	−0.114	0.536	−0.179	0.165	−0.289	−0.255	−0.077	0.010	−0.115
EI07	0.509	−0.156	0.043	−0.243	−0.218	−0.115	−0.162	−0.261	0.536	−0.164	0.045	−0.256	−0.230	−0.121	−0.171	−0.275
SI04	0.522	0.000	−0.260	−0.083	−0.007	−0.036	0.123	0.094	0.526	0.000	−0.262	−0.084	−0.008	−0.036	0.124	0.095
SI03	0.520	−0.073	−0.035	0.303	−0.359	0.243	0.091	−0.016	0.524	−0.074	−0.035	0.306	−0.362	0.245	0.091	−0.016
EI02	0.396	−0.147	0.017	0.014	0.104	−0.061	0.136	−0.047	0.509	−0.189	0.022	0.018	0.134	−0.079	0.175	−0.061
EI10	0.492	−0.038	−0.044	0.366	0.197	0.099	−0.158	−0.065	0.501	−0.038	−0.044	0.372	0.201	0.101	−0.161	−0.066
SI08	0.443	−0.211	−0.431	−0.105	0.063	0.188	−0.145	0.278	0.459	−0.218	−0.447	−0.109	0.065	0.195	−0.151	0.288
SI01	0.358	−0.114	0.078	0.026	0.046	−0.093	0.122	0.087	0.456	−0.146	0.100	0.033	0.059	−0.119	0.155	0.111
EI06	0.428	−0.231	−0.216	0.146	0.121	0.233	−0.183	−0.114	0.449	−0.242	−0.227	0.153	0.127	0.245	−0.192	−0.120
SI09	0.409	−0.079	−0.204	0.386	−0.213	−0.020	−0.095	0.127	0.429	−0.083	−0.214	0.405	−0.224	−0.021	−0.100	0.133

SI12	0.386	−0.091	−0.006	−0.289	0.087	0.337	−0.163	0.329	0.413	−0.097	−0.006	−0.309	0.093	0.361	−0.175	0.352
SI07	0.203	0.860	0.493	0.083	−0.485	0.017	−0.062	0.117	0.165	0.700	0.401	0.068	−0.395	0.014	−0.051	0.095
EI08	0.281	0.503	−0.061	0.236	0.372	0.236	0.493	−0.039	0.265	0.474	−0.058	0.222	0.351	0.223	0.465	−0.037
SI06	0.493	0.783	−0.809	−0.160	0.075	−0.397	−0.155	−0.043	0.363	0.577	−0.596	−0.118	0.055	−0.293	−0.114	−0.032
SI13	0.203	0.334	0.377	0.256	0.256	−0.183	−0.159	0.169	0.210	0.345	0.390	0.265	0.265	−0.189	−0.165	0.175
SI10	0.313	0.474	0.099	−0.552	0.027	0.476	0.183	−0.020	0.283	0.429	0.090	−0.501	0.024	0.431	0.166	−0.018
SI11	0.322	0.137	0.228	0.367	0.056	−0.209	−0.309	0.218	0.341	0.145	0.242	0.390	0.060	−0.222	−0.328	0.232
EI12	0.414	0.109	0.214	0.182	0.438	0.126	0.033	−0.255	0.425	0.111	0.219	0.187	0.449	0.130	0.033	−0.262
SI14	0.380	0.009	0.356	−0.320	0.281	0.109	−0.397	0.151	0.391	0.009	0.365	−0.329	0.289	0.112	−0.408	0.156
EI04	0.385	0.065	−0.044	0.197	−0.041	0.292	−0.253	−0.545	0.391	0.066	−0.044	0.200	−0.041	0.297	−0.257	−0.554

Extraction method: principal component analysis

[a]Eight components extracted

Table 22.9 Summary of item statistics

	Mean	Minimum	Maximum	Range	Maximum/minimum	Variance	No. of items
Inter-item covariances	0.153	−0.219	0.497	0.716	−2.272	0.009	28

Table 22.10 ANOVA with Friedman's test and Tukey's test for non-additivity

Groups			Sum of squares	df	Mean square	Friedman's chi-square	Sig
Between people			2403.374	473	5.081		
Within people	Between items		887.133	27	32.857	41.505	0.000
	Residual	Non-additivity	2.089[a]	1	2.089	2.640	0.104
		Balance	10,107.921	12,770	0.792		
		Total	10,110.010	12,771	0.792		
	Total		10,997.143	12,798	0.859		
Total			13,400.517	13,271	1.010		

Grand mean = 2.28

[a]Tukey's estimate of power to which observations must be raised to achieve additivity = 1.260

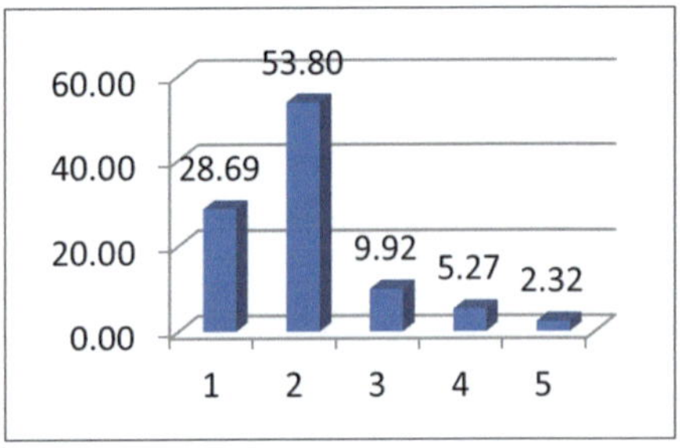

Fig. 22.5 No development at all (DT1)

cultural or educational showed normal distribution. Development choice constructs like only historic or only recreational use or only water-dependent use like water-based transport, industries, commercial fishing, etc. show raised cosine distribution of more rounded peak and thinner tails. Refer Figs. 22.5, 22.6, 22.7, 22.8, 22.9, 22.10, 22.11 and 22.12 for the above descriptions. Table 22.11 presents the descriptive statistics of type of development supported or preferred by the respondents for the Sabarmati Riverfront.

Majority of the respondents are opposed to single-use development with almost 82 % opposing the no development option and almost 74 % preferring the mixed-use development as the most appropriate model of development. Among single-use development type, only residential riverfront development is highly opposed followed by only commercial, only working and only recreational development, respectively. Table 22.12 presents component matrix of type of riverfront development supported

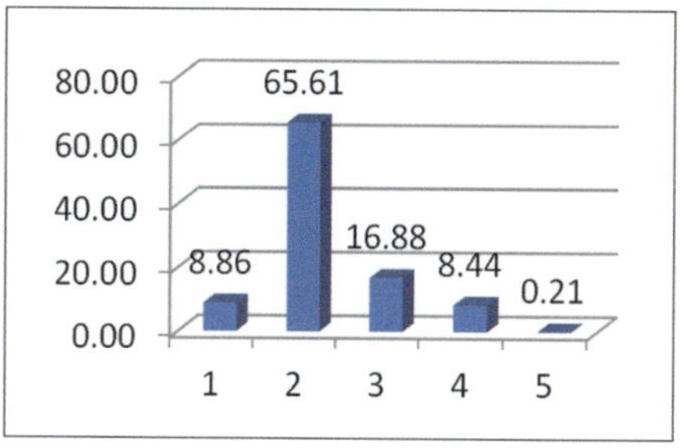

Fig. 22.6 Support or oppose only commercial riverfront development (DT2)

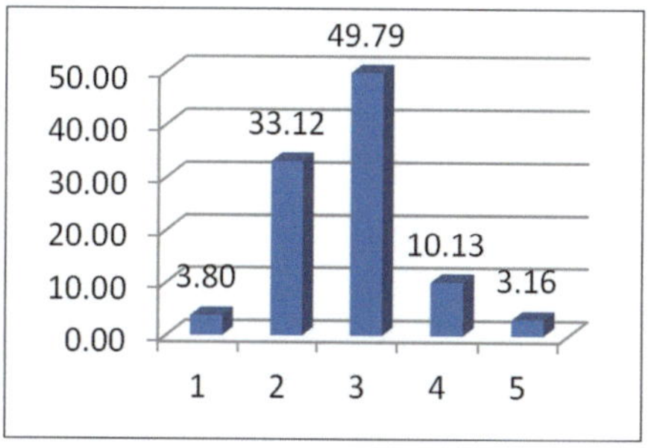

Fig. 22.7 Support or oppose only cultural, educational and environmental riverfront development (DT3)

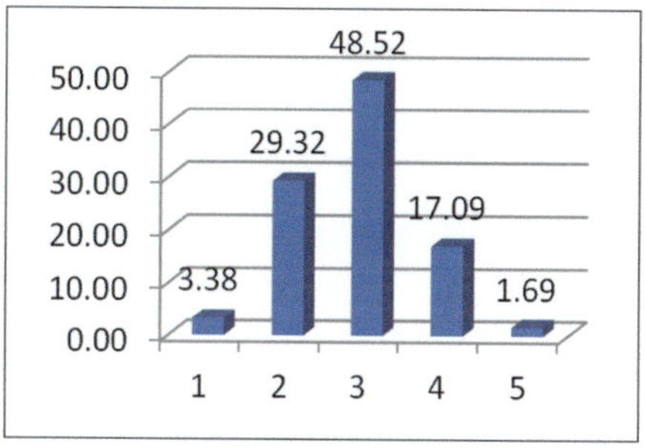

Fig. 22.8 Support or oppose only historic riverfront development (DT4)

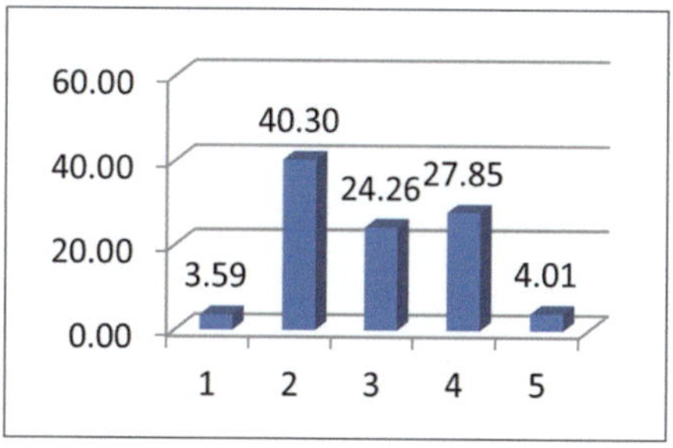

Fig. 22.9 Support or oppose only recreational riverfront development (DT5)

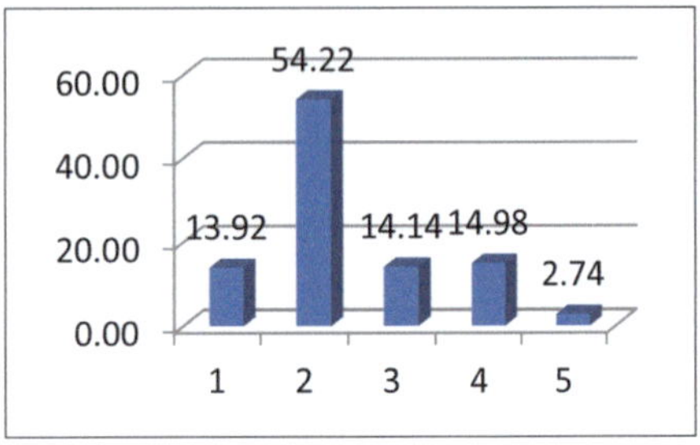

Fig. 22.10 Support or oppose only working riverfront development (DT6)

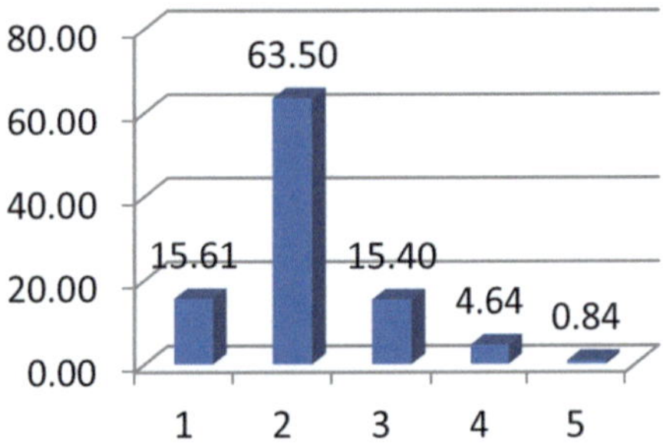

Fig. 22.11 Support or oppose only residential riverfront development (DT7)

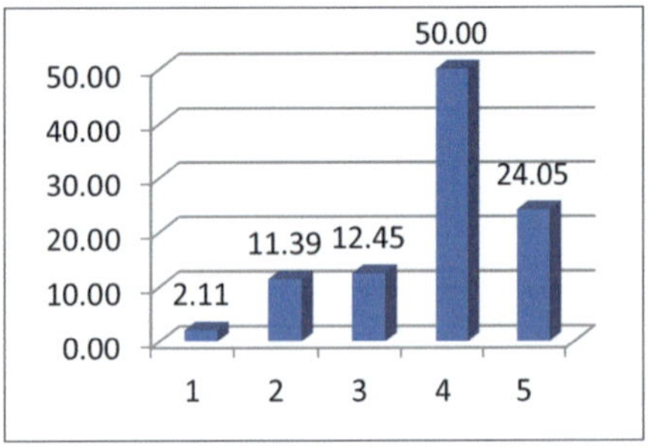

Fig. 22.12 Support or oppose mixed-use riverfront development (DT8)

Table 22.11 Descriptive statistics of type of riverfront development supported by the residents

Variable	Mean	SD	Skewness	Kurtosis
DT1	1.988	0.899	1.236	1.891
DT2	2.255	0.742	0.926	0.929
DT3	2.757	0.808	0.397	0.581
DT4	2.844	0.802	0.092	−0.057
DT5	2.884	0.987	0.261	−0.095
DT6	2.384	0.991	0.783	−0.016
DT7	2.116	0.748	0.965	1.861
DT8	3.825	0.991	−0.874	0.269

Table 22.12 Component matrix of type of riverfront development supported by local residents

Component matrix[a]								
	Raw				Rescaled			
	Component				Component			
Variable	1	2	3	4	1	2	3	4
DT1	0.497	−0.011	0.430	−0.038	0.553	−0.012	0.479	−0.043
DT2	−0.038	0.389	0.153	−0.071	−0.051	0.525	0.207	−0.095
DT3	−0.246	0.220	0.540	−0.022	−0.304	0.273	0.667	−0.027
DT4	−0.164	0.130	0.523	−0.069	−0.204	0.162	0.652	−0.086
DT5	−0.714	0.034	0.349	−0.050	−0.723	0.034	0.353	−0.051
DT6	0.390	0.775	−0.098	−0.264	0.394	0.782	−0.099	−0.267
DT8	−0.449	0.563	−0.383	0.406	−0.453	0.568	−0.386	0.410

Extraction method: principal component analysis
[a]Four components extracted

by local residents, analysis of which shows there are four components cumulatively explaining 60 % of the variance. The first component comprises of the development choice no development at all on the riverfront (DT1) explaining 16.41 % of variance. The second component is made up of single-use development options of only commercial riverfront development (DT2), only working riverfront development (DT6) and only residential riverfront development (DT7), respectively, which cumulatively explains 15.3 % of the variance.

The third component comprises of only cultural, educational and environmental riverfront (DT3) and only historic riverfront DT4 explaining 15.08 % of the variance. All the three components are opposing in nature. However the fourth component comprises of support for mixed-use riverfront development (DT8) explaining 12.98 % of the variance.

22.3.8 *Proposed Structural Model of Perceived Impact Dimensions of Urban Riverfront Development by Local Residents*

Among all the 12 perceived economic impact constructs, only eight and, of the total 14 perceived social impact constructs, only six are identified as strong components contributing to overall impact perception dimension of riverfront development. The rest of the impact dimensions are not perceived in a significant manner.

The proposed model (refer Fig. 22.13) is therefore identified by grouping eight economic constructs (EI03, EI11, EI05, EI09, EI07, EI02, EI10) and four social constructs (SI05, SI04, SI03, SI02) that constitute component one represented in the model as C1. Component 2 represented as C2 in the model constitutes of one economic construct (EI08) and two social constructs (SI07, SI06). Among component 1 (C1), the economic construct that riverfront development will attract more investment

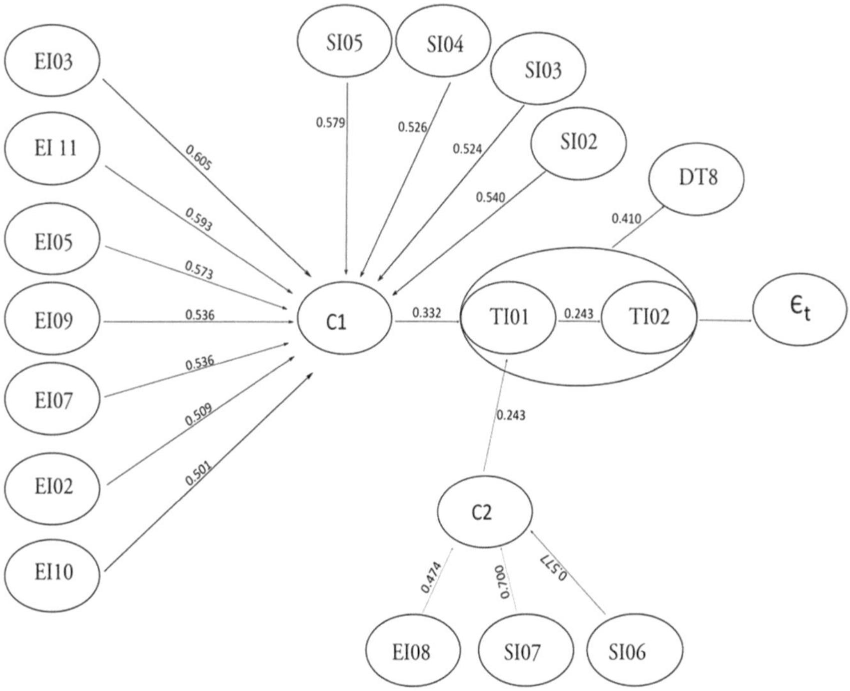

Fig. 22.13 Path diagram of the proposed simple structural model

to your community is strongly perceived, and the least perceived construct is that the riverfront development will increase real estate tax revenues. Among perceived social constructs, the construct that riverfront development will create and promote a more positive image of the city is strongly perceived. The least perceived social construct is that riverfront development will create new public spaces and new places to visit, admire and enjoy, which is in conflict with one of the objectives for the Sabarmati Riverfront Development Project.

In component 2 (C2), the economic construct that costs of developing riverfront facilities are too much of a burden on government/public sector agencies is significantly perceived. Social constructs that riverfront development is a misuse/waste of taxpayers' money and that riverfront development will displace low-income and ethnically different populations from the riverfront area are strongly perceived for component 2. Both of these impact dimensions are of major concerns for the project already and the reasons which have contributed much for its delay. The structure also reflects that 74 % of the respondents support mixed-use development.

22.4 Conclusion

This study in context to Sabarmati Riverfront Development Project, Ahmedabad, highlights that the local residents have clear perception that the riverfront development has a positive overall impact on the city. However with regard to the social and economic impact of riverfront development, the residents have strong perception about some constructs and weak for some. Approaches need to be considered to strengthen the positive impact components which are perceived strongly and which contribute significantly to the overall perceived impact and take measures to nullify the strongly perceived negative impact components in order to ensure continued support from the local residents and the community. Survey input-based focused group formation of residents can be considered to ensure participation of the community in the riverfront development process. Regarding the type of development that should come along the riverfront, the perception of the local residents about the manifestation of development is not very clear. They are not sure about the type of riverfront development that should come up on the riverfront. For this, SRFDCL, which is entrusted with the task of developing the riverfront, should have involved the local residents in the riverfront planning and development process from the very inception stage. Local residents should have been made aware of the different possible models of development that has been considered for such areas the world over and also identify their own priorities and needs about such developments. This would have then helped them in ascertaining the suitable model of development for their riverfront.

Similar replicating studies can be adopted for other riverfront cities undergoing or considering development of their riverfront areas, to identify the perception of their residents about their riverfronts and to ensure sustainable, smooth, uninterrupted and inclusive and people-centric development, benefitting the city at large.

References

Bentler PM (2005) EQS 6 structural equations program manual. Multivariate Software, Encino (www.mvsoft.com)

Breen A, Rigby D (1994) Waterfronts: cities reclaim their edge. McGraw-Hill, New York

Breen A, Rigby D (1996) The new waterfront: a worldwide urban success story. McGraw-Hill, New York

Bruttomesso R (1993) Working on the water's edge. In: Bruttomesso R (ed) Waterfronts- a new frontier for cities on water. International Center Cities on Water, Venice, pp 10–11

Byrne BM (2010) Structural equation modelling with AMOS: basic concepts, applications and programming. Taylor and Francis, New York

Craig-Smith SJ, Fagence M (eds) (1995) Recreation and tourism as a catalyst for urban waterfront redevelopment. Praeger, Westport

Hair JF, Anderson RE, Tatham RL, Black WC (2008) Multivariate data analysis, 7th edn. Prentice Hall Publisher, Upper Saddle River, New Jersey

Hoyle BS, Pinder DA (1992a) European port cities in transition. Belhaven, London

Hoyle BS, Pinder DA (1992b) Cities and the sea: change and development in contemporary Europe. In: European Port Cities in transition, London, Belhaven

Hoyle B (1997) Review of the new waterfront: a worldwide urban success story. In: Breen A, Rigby D (eds). J Trans Geogr 5(4):304–306

Hoyle B (2000) Global and local change on the port-city waterfront. Geogrl Rev 90(3):395–417

Malone P (1996) City, capital, and water. Routledge, London

Martire A (2009) "Waterfront retrieved," Buenos Aires' contrasting leisure experience. In Maciocco G, Serreli S (eds) Enhancing the city: new perspectives for tourism and leisure. Springer 6:245–274

Norcliffe G, Basset K, Hoare T (1996) The emergence of postmodernism on the urban waterfront: geographical perspectives on changing relationships. J Transp Geogr 4(2):123

Index

F. Seta et al. (eds.), *Understanding Built Environment*, Springer Transactions in Civil and Environmental Engineering, DOI 10.1007/978-981-10-2138-1

MIX
Papier aus verantwortungsvollen Quellen
Paper from responsible sources
FSC® C105338

If you have any concerns about our products,
you can contact us on
ProductSafety@springernature.com

In case Publisher is established outside the EU,
the EU authorized representative is:
Springer Nature Customer Service Center GmbH
Europaplatz 3, 69115 Heidelberg, Germany

Printed by Libri Plureos GmbH
in Hamburg, Germany